Internal Time

Internal Time

Chronotypes, Social Jet Lag, and Why You're So Tired

Till Roenneberg

Harvard University Press · Cambridge, Massachusetts · London, England · 2012

First published in German translation as *Wie wir ticken: Die Bedeutung der inneren Uhr für unser Leben,* copyright © 2010 by DuMont Buchverlag, Cologne (Germany)

Many of the designations used by manufacturers and sellers to distinguish their products are claimed as trademarks. Where those designations appear in this book and Harvard University Press was aware of a trademark claim, then the designations have been printed in initial capital letters.

Library of Congress Cataloging-in-Publication Data

Roenneberg, Till.
[Wie wir ticken. English]
Internal time : chronotypes, social jet lag, and why you're so tired /
Till Roenneberg.
p.   cm.
Subtitle of original German text: Die Bedeutung der inneren Uhr
für unser Leben.
Includes bibliographical references and index.
ISBN 978-0-674-06585-7 (alk. paper)
1. Chronology.   2. Biological rhythms.   I. Title.
QP84.6.R64 2012
612'.022—dc23   2011050230

For Casper, Pauline, and Flora

Contents

Internal Time

This book is about clocks. Not about those you can buy, wear, or hang on a wall, but about the clock that ticks away in your body. The body clock is not a new invention in the long line of evolution. You share your ability to internally keep track of the time of day with practically every other creature on earth, from other mammals right down to those organisms that exist only as a single cell. This means that an internal biological clock must be extremely important for life on this planet. Living without or against the biological clock would mean premature death by predation or starvation for most animals in their natural environment. As I will elaborate in this book, living against our biological clock also jeopardizes our health and well-being. In modern society, we rarely live in synchrony with our body clock. Some of us travel fast across many time zones, and others (about 20 percent of the working population in industrialized societies) have to work in shifts. If you have ever suffered from jet lag, you know how strongly we are affected when our body clock is out of synch with social time. But even if you don't work in shifts and never travel across time zones by airplane, you can still suffer from a chronic type of jet lag, which we call *social jet lag.*

A book about clocks is evidently also about time, yet the time of the body clock is not necessarily the same as the social time that we use in everyday life to be on time for work, travel, appointments, or the evening news. Social time is the time people live by. In the nineteenth century, social time was local sun time: noon was when the sun had reached its highest point. This rather sensible long-standing convention about time was eventually challenged by trains: they could transport humans over long distances within a few hours, rendering local sun time impractical because passengers had to reset their clocks at almost every station. States and countries, therefore, adopted a universal system in 1884, subdividing the world into twenty-four time zones that all refer to a zero line, the longitudinal meridian that crosses the observatory in Greenwich, near London. Theoretically, social time could adopt any time frame, as long as everyone used the same one. The entire territory of mainland China, for example, is covered by a single time zone based on Beijing time. This book will tell you how these different time systems interact: *sun time, social time,* and your very individual *internal time.*

Your internal time is produced by your own body clock. It varies from individual to individual just as body height, eye color, or personality varies, and it interacts with sun time and social time. In spite of internal time being probably the most important to our health and well-being—more important than sun time and certainly more important than social time—it has been thoroughly neglected. Every day, we are awake for approximately sixteen hours until we render the control of our movements, thoughts, and desires to a state resembling unconsciousness, which we call sleep. These daily changes are so utterly obvious that the underlying biological mechanism remained unexplored for centuries. In synchrony with the rise and fall of the sun, animals awaken and sleep, plants open and close their blossoms, and plankton travel up and down the water column. All these rhythms are controlled by a biological clock that represents the twenty-four-hour day of our planet. The alterations between sleep and wake are not simply two states of our existence that are being flipped like a coin once a day. They are reflections of a continuous change in all our bodily functions; a change that involves both turning genes on and off and continuously changing the cocktail of hormones and transmitters in all our tissues.

I have been studying the biological mechanism underlying our body clock for decades in very different creatures ranging from single cells and simple fungi to humans. These studies were conducted either in the laboratory, where we try to control all environmental factors (such as light, temperature, and food), or in the real world—for example, in factories, where we measure different variables over the course of the day, or in the world at large, by simply asking normal people when they do what. My initial reason for studying the body clock was autobiographical and almost by chance. One of the pioneers of the field, Jürgen Aschoff, was the director of a research institute in the Bavarian countryside. Jürgen and his wife, Hilde, had six children, who attended the same school that I did. I became friends with all of them, despite age differences. The Aschoff family lived in a beautiful castle ("the Schloss") on the slope of a hill in Andechs, a village near the Ammersee, one of the larger Bavarian lakes. Andechs was a long way out in the countryside with almost no commuting possibilities. The Aschoff parents, therefore, allowed their children to invite friends to stay whenever they wanted, even overnight. To be part of a large crowd of youngsters who were all extremely interesting and interested was a lot of fun, and I stayed at the Schloss as often as I could. I also got on extremely well with the professor himself and became more and more interested in the science that he and his colleagues pursued.

When I was about seventeen, I began to work as a part-time assistant at Aschoff's institute during most of my school holidays. It allowed me to combine the fun of learning science, spending time with fascinating people, and earning some money—a dream situation. There were always many visitors: friends of the parents; friends of the children; but also lots of scientists, some of them internationally famous, who seemed to be engaged in endless scientific discussions. I had always loved science, but the atmosphere in Andechs gradually turned me into a junkie—this was exactly the life I wanted to lead.

Although I probably had assimilated more knowledge about body clocks by the time I entered university than most students do by the time they graduate, I started to study physics, in my view the most fundamental science of all. But I soon realized that the true aim of my interests concerned humans and that physics just didn't get me any closer to understanding them. So I switched to medicine. But once again, I felt that this discipline would not be the right vehicle for my curiosity. Although I wanted to know everything about humans, the focus of my interests wasn't on helping or curing them. I began to appreciate that I could only start to learn something about humans if I understood more about evolution, genetics, biochemistry, comparative physiology, and ecology, but none of these subjects was taught in any depth in medical school.

I finally ended up studying biology. I enjoyed discussions with other students and scientists much more than attending lectures, and I read lots of papers and books that covered many more issues than those taught in a degree program. I felt that my learning only began when I started working in the lab, collecting experimental data and then trying to make sense of them. The attempts at making sense of data always were my utmost joy in science—and still are. I think that this joy, and the ways and methods I use to approach the mysterious world of data, go back to my intensive interactions with Jürgen Aschoff when I was at a very impressionable age.

After a detour into photobiology, neurophysiology, and brain research that lasted for many years, I finally returned to the field that investigates biological clocks (chronobiology) as a postdoctoral fellow. I spent my first postdoc time back in Andechs, working with Jürgen Aschoff—not as a student but as a (somewhat) fledged scientist. I remained a colleague and friend of "the old one," as his family and his close friends called him (although he never ceased to be my mentor), until he died in October 1998. After two years studying annual rhythms in humans with Aschoff, I was eager to learn more about how biological clocks work in cells, how they generate an internal day with the help of molecules. So I decided to work with another pioneer of chronobiology, Woody Hastings, a Harvard professor. I was part of his team in Cambridge, Massachusetts, for almost four years and kept going back during the summer for many more years. Returning to Germany, I found a rocky academic landscape for someone who simultaneously studied humans and single-cell algae, someone who was more interested in the investigation of concepts, such as the biological clock, than in staying within the boundaries of the little boxes created by our disciplines and faculties. Where did I belong according to orderly German academic criteria—botany, zoology, ecology, anthropology, or medicine? I ended up in the medical faculty, specifically in the department of medical psychology. Ernst Pöppel, the chair of the department, provided this scientific home, where I still reside. He was one of the few also interested in concepts (especially those regarding time), and less focused on the specific model organisms that were used to study them.

Over the years of studying the biological clock, I began to realize that what clock researchers were discovering had an enormous significance for our everyday lives. I noticed that people were fascinated and eager to learn about the science behind circumstances of their daily existence, some of which they had never considered before. Once enlightened, they started to understand themselves (and others) much better, began to appreciate their own individual time, and were suddenly relieved of the weight of prejudice ridiculing their temporal habits: for example, being called lazy if you don't wake up fresh as a daisy by seven o'clock in the morning; or being called a boring person only because you don't enjoy going out with friends after ten at night.

In this book I tell a story about internal time, or rather many little stories covering different aspects of our body clock. Each of the twenty-four chapters—I'd be telling lies if I told you I chose to write twenty-four chapters by chance—has two sections, the case and the background. In the case, I lay out a story about internal time that the later background information explicates. In many of these cases I will manipulate the facts for the sake of a good story, but in all, the data concerning chronobiology are accurate. For example, no one knows exactly what passed through a certain eighteenth-century scientist's head when he came terribly close to discovering the phenomenon of internal time (which then lay idle for almost a quarter of a millennium), but, based on what he wrote, I use my imagination to get the facts across to you. Some cases describe relatively recent discoveries made by contemporary scientists. They are written to draw your attention to a question and to help you imagine how a discovery could have taken place. Although the scientific facts of the discovery itself are historically correct, other details, such as names and places, may be purely fictional.

With the case stories, I want to raise your curiosity and pique your urge to reason. If you are puzzled by a case, first try to identify what you don't understand, and then consider what parts of the story are reflected in your own life. The second section of each chapter describes the facts underlying the case story in detail. It should answer most of your questions and may help you relate what you read to your own temporal life.

The use of cases is part of the philosophy of problem-based learning. Its aim is to focus the mind on a problem without employing jargon or excessive scientific explanation. Problem-based learning is often applied in university education, particularly in medical, law, or business schools. Students, often working in groups, are asked to identify all the facts in the case and then work out the background behind the story with the help of textbooks, the internet, lectures, and specialists. The best part of problem-based learning is being forced to confront an everyday problem that is completely puzzling, at least at first. The drawback of traditional learning has always been the dissociation between the theory and its application. "Why do we have to learn this?" is probably one of the most frequent and justified questions teachers hear.

To understand the biology of the body clock does require some knowledge of biology. I have done my best to keep the biological explanations accessible for everyone. More detailed explanations can be found in the endnotes. Whether you are interested in just the case studies, the background information, or the more detailed scientific explanations in the endnotes, I hope you will gain a thorough understanding of internal time, *your* time. I want you to appreciate how important this concept is for living well. I hope you have as much fun reading this book as I had writing it—fun is the best way to understanding and allows us to remember new information without too much effort.

1

Ann woke to a hard and persistent knock on her bedroom door. After staying in bed as long as possible, she wrapped herself into a thick bathrobe, put on warm socks, and stumbled to the bathroom to brush her teeth. She didn't say "good morning" to her father and didn't expect any form of greeting from him, either. If she hadn't pushed him grumpily away from the basin to reach the faucet, one would have thought that neither was aware of the other. It was a school morning, shortly before the Christmas break. Ann was, as usual, far behind schedule and, like her father, Jim, began the new day in a state of semi-consciousness. Before Jim had school-aged children he used soap and a razor, which he thought produced a much cleaner shave, but now that he was forced into this early routine, he tolerated the noise of an electric razor, having cut himself too often. His wife, Helen, was already downstairs making breakfast together with their son, Toby.

In contrast to father and daughter, who went about their morning routines in silence, Toby and his mother were chatting with each other like a pair of canaries. Toby was telling her about his field trip to the dinosaur exhibit and got quite carried away when enumerating the different raptors he had seen. While his mother was preparing sandwiches for the children's school breaks, Toby set the table, but soon got distracted by the back of the cereal box, which he had just placed in front of his bowl. He read all about the new dinosaur collection that would be coming out next year—one in each package.

He decided to eat more cornflakes in the future, at least two bowls every morning.

Helen always put extra effort into the sandwiches for her first-born. She wanted to make sure that her daughter would eat them because she usually left the house with nothing more than a cup of tea in her stomach. By the time Ann had crossed the threshold into puberty, she had stopped eating anything before leaving the house, and Helen had fought—and lost—endless battles about "proper breakfasts." It was actually Jim who had put an end to this struggle: "Make her a good sandwich with her favorite stuff on it and she will eat it in school as soon as she is hungry." Of course, no parent can ever be sure what really happens to lunches from home, but the fact that Ann uttered requests from time to time for variations in the narrow theme of her favorite foods encouraged Helen.

At around seven o'clock Jim joined the talkative pair, kissed Toby on his way through the kitchen and then Helen, who handed him a big mug of coffee. The three sat down at the breakfast table, and Helen, as usual, yelled, "Hurry up, Ann, the bus will be here in twenty minutes!" They were lucky because the bus stop was right in front of their house, and Ann made the best of this short distance by coming downstairs only a couple of minutes before it arrived. Helen's primordial cry was a remnant of her "proper breakfasts" fight.

At last, Ann did join the others, slowly sitting down at the table to sip her tea. Toby continued his lively dinosaurial narrative, mainly addressing his mother. He only approached his sister at breakfast if he was in a mischievous mood. She was an easy victim, unable to muster resistance, although later in the day she usually did get her revenge. Helen was half listening to Toby and half concentrating on planning the day, making her to-do list, and giving Jim or Ann short instructions. The chances that Helen could have a real conversation with her husband in the morning were greater in summer, when daylight flooded the kitchen and they occasionally had breakfast on the terrace. Now that the sun came up after the children were in school, all of them were more subdued than usual—even Toby and his mother. Helen was thinking of the PTA meeting scheduled for that evening and decided to ask Jim to go—he was much more up to it at that time of day, and she could go to bed early.

Ann brooded about the school day ahead of her. Why math in the first period? Why not art or history? She was actually quite good at math, but she needed at least half a functioning brain to solve mathematical problems, and most of her brain certainly didn't wake up before ten o'clock or even later, no matter how early or late she got up. When Ann left the room to get her coat, Jim was able to produce the first smile of the day when he read the back of his daughter's tee shirt: "Early to rise and early to bed makes a bird healthy, wealthy, and dead."[1]

This first case represents—with minor variations—the morning routine of millions of households across the globe. In that sense, it is almost trivial. But apparent trivialities will play an important role in this book. How easily do we wake up in the morning, and why? Jim, Helen, Ann, and Toby seem to be worlds apart at this early hour: Helen and Toby feel fresh, and Jim and Ann feel woolly. The case is rich in information that you might have absorbed without even noticing, but it also triggers many questions: Is this a sex or gender issue?[2] Does age play a role? Does the ability to wake up depend on the time of day? Are different wake-up types also different fall-asleep types? What do eating habits have to do with these different wake-up types? Does performance in such different activities as math and art depend on the time of day? The wake-up type? Some of these questions can be answered by the story itself.

Jim, Helen, Ann, and Toby's morning touches upon many different aspects of temporal life and biological clocks. We are told when the story takes place (an early weekday morning shortly before Christmas), but we need to know where the family lives. Our conclusions will differ radically if the family lives in South Africa, Peru, or Australia rather than Europe, Japan, or the United States. But the story gives several hints to indicate that the family lives somewhere in North America or northern Europe: if it is nearly Christmas and dark outside, the family must live in the northern hemisphere. The size of the house and family and what they eat may provide other hints.

It would also be helpful to know the approximate ages of the family members. Let's start with an educated guess: Toby appears to be younger than Ann, and Jim is probably older than Helen. Toby must be about six or seven because he is still young enough to be fascinated by dinosaurs but old enough to read the back of the cereal box. Ann, firstborn, has entered puberty but still lives at home, so let's put her at fourteen.[3] Helen and Jim, their parents, therefore might be in their early forties. Theoretically, Helen could be much younger and Jim much older, but that is somewhat beside the point.

Now let's turn to the central theme—the different wake-up types. Father and daughter are not exactly communicative. Is Ann mad at her father or does her grumpy behavior merely reflect her resentment of having to get up at that ungodly hour? We could ask the same questions about Jim. By contrast, Helen and Toby are in the best of moods, already fully awake and active. Is this just the normal contrast between teenagers and children, or could these behavioral differences be—at least partly—accounted for by different wake-up types? Being a certain wake-up type is obviously not a simple matter of age, sex, or gender since both Helen and Toby are more awake than either Jim or Ann.

Once again, the story answers our questions. Before Jim had children at school, he felt vigilant enough to give himself a clean shave. Ann is apparently quite good at math after ten o'clock. Different wake-up types are apparently also different fall-asleep types. Helen is fresh in the morning but feels sleepy quite early in the evening. Although Jim has to get up on weekdays at approximately the same time as Helen does, he is dead tired in the morning but still quite alert in the evening—the PTA meeting will, therefore, be his task.

We know from experience—starting in our own families—that individuals possess different timing types, or *chronotypes*.[4] In many cultures and languages, chronotypes are often named after birds— early birds and late birds.[5] The common usage of *larks* and *owls* suggests that we are dealing with two categories. A Danish researcher has recently coined them, less poetically, A and B types, supporting the notion of two categories. However, the attempt to categorize any population of living beings into two categories is rarely correct. In general, human qualities, including chronotype, almost never fall strictly into two simple categories.

My colleagues and I have investigated human chronotypes for many years by asking thousands of people about their sleep habits with the help of a questionnaire.[6] We use the answers to these questions to define a person's chronotype. Defining the timing of an event is not necessarily straightforward. "When did you hear the shot?"; "When is high tide?"; or "When did the sun rise?" are easy questions because they concern clearly definable events. "When do you usually sleep?" is more complicated because we usually sleep for seven to eight hours. We therefore ask "When do you usually fall asleep?" or "When do you usually wake up?" But the answers to even these questions are difficult to translate into a person's chronotype. Let's suppose that Person A sleeps from 10 P.M. to 6 A.M., Person B from 10 P.M. to 8 A.M., and Person C from midnight to 6 A.M. If chronotype were defined by sleep onset, A and B would be the same type. If one defined it by sleep end, A and C would fall into the same category. The difficulty arises because sleep has (at least) two different and independent qualities: sleep *timing* and sleep *duration.*

It turns out that the midpoint of sleep is best for defining a person's chronotype and also solves the problems described above. The calculation of midsleep is easy: if a person usually falls asleep at midnight and usually wakes up at eight, then his usual midsleep is 4 A.M. All these sleep times should represent what is done daily, not what is the exception, such as a party or a late night at work. Midsleep of Person A would be 2 A.M., that of Person B would be an hour later at 3 A.M., and Person C would have the same midsleep as B but would sleep four fewer hours, going to bed two hours later and waking up two hours earlier.

Our large database allows us to investigate the epidemiology of sleep behavior in different populations worldwide.[7] The figure shows the distribution of midsleep in Central Europe (containing the answers from approximately 100,000 participants, predominantly Germans).[8] The distribution is almost a perfect bell shape, although late types are slightly more numerous than early types.[9] Categories like larks and owls, A and B people, misrepresent the continuous distribution of chronotypes as much as dwarves and giants misrepresent the distribution of body height. These opposites simply label the extreme types at both ends of distributions, which are extremely rare.

We base the first assessment of an individual's chronotype on her sleep behavior on free days, when it is not dominated by work or school times but rather by individual preference, by a body clock that dictates her *internal time.* Slightly over 14 percent of the population (represented by our database) fall into a midsleep category of 4:30 to

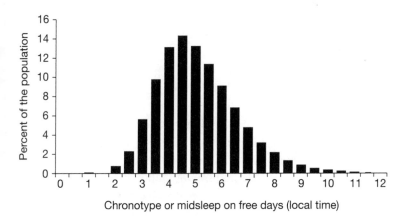

The distribution of midsleep in Central Europe.

5:00 A.M. Presuming a sleep duration of eight hours (to make things simple), individuals in this category go to bed on free days between half past midnight and 1:00 and wake up between 8:30 and 9:00 A.M. The midsleep times (on free days) of over 60 percent of the population fall between 3:30 and 5:30 A.M., but only less than half a percent between 1:30 and 2:00. These extreme larks begin their sleep on free days between 9:30 P.M. and 10:00 and wake up voluntarily between 5:30 A.M. and 6:00 (again assuming an eight-hour sleep duration). There are even more extreme early chronotypes, but their number is so low that the corresponding vertical bars are too small to be detectable in this graph. On the late side of the chronotype distribution, about four percent fall asleep between 3:00 and 3:30 A.M., and many more people sleep even later.

So far, we have identified chronotypes primarily based on sleep habits, but it will become clear throughout this book that chronotype means much more. The internal timing of our body clock, of which sleep habits are just one aspect, dominates all functions in our body, ranging from genes being activated at certain times, to changes in body temperature and hormonal cocktail, right up to cognitive functions like the ability to do math. Magazine articles about the body clock often tell their readers when to do what: when to go to the dentist because the pain sensation is minimal; when to exercise because the training effect is greatest; when the best times are to do math or write poems or even when to make love. I am amazed and happy that the media pick up on the importance of our biological clock but also a bit worried that many of these articles approach this phenomenon naively. Some of them advise their readers, for example, to have sex in the morning. This advice is merely based on the fact that the male sex hormone testosterone peaks in the early morning hours. This may lead to an increased sex urge in men and may even ensure better male "performance." But does this necessarily mean that sex is optimal for both partners at this time of day? Our biology concerns all aspects of our being, including our psychology and feelings, from pain to pleasure, and all these facets put together are more complex than the concentration of a single hormone, however important testosterone may be for sex. Thus, the media should use care when giving advice. This holds especially in light of what we have learned in this chapter: different people can have very different chronotypes with the extremes being up to twelve hours apart. So even if a piece of advice about when to go to the dentist was valid and useful, it would only be of limited help because an individual's internal time could theoretically be as many as six hours earlier or later than the advised time.

We are capable of different things at different times of the day. Many aspects of performance show daily fluctuations, and their highs and lows depend on chronotype. Not only performance but many other aspects of life vary according to the time of day. If we wake up in the middle of the night, we don't cook ourselves a meal—we are not hungry. So why should Ann be hungry just because she is being woken every school day at around her internal midnight? The quotation on Ann's tee shirt questions why the early hours of the day have such a good reputation, with all those proverbs glorifying the early risers. Good question!

2

The sun had just risen above the horizon when the farmer walked along a country path toward his field. He cheerfully greeted a man he encountered halfway between the village and his destination. He thought to himself, "Must be a decent bloke to be up so early."

———

The postman rang the doorbell at half past ten. If he hadn't heard noises from within the apartment, he would have left long before. He rang the bell once more—longer and harder—and then heard a rather grumpy voice announcing the imminent appearance of its owner. "Lazy bugger," thought the postman when the young man appeared in a robe and with sleep-tousled hair.

———

The journalist had waited for quite some time on the phone. She was finally put through to the scientist. "Thank you very much, professor, for making the time to answer my questions. I wanted to interview you about early birds and long sleepers." The researcher looked toward the heavens and suppressed a long sigh, knowing that this interview would involve debunking many myths and explaining many basic principles.

The moral backbone of practically every culture declares early risers as good and late risers as bad people. God rewards the early risers; they are more effective ("an hour in the morning is worth two in the evening") and will be wealthier than the "lazy" rest of the population. The larks are declared the successful and productive members of society while owls are, at best, extroverted artists and intellectuals, or at worst, people who engage in dark arts and exert evil powers. Folk wisdom praising early risers can be found in many places around the globe: "The early bird catches the worm," "Zǎoqǐ de niǎo er yǒu chóng chī" (United States and China); "Morgenstund hat Gold im Mund," "De ochtendstond heeft goud in de mond" ("The morning hour has gold in its mouth"—Germany and the Netherlands); "À qui se lève matin, Dieu aide et prête la main" ("God helps those who rise in the morning and gives them a hand"—France); "Chi dorme non piglia pesci" ("Those who sleep won't catch fish"—Italy); "A quien madruga Dios le ayuda" ("God helps those who get up early"—Spain); "Kto poran'she vstaët, tot gribki sebe berët; a sonlivyi i lenivyi idut posle za krapivoi" ("The early riser gathers mushrooms, the sleepy and lazy one goes later for the nettles"—Russia).

Many of them are mere translations or variations of the early-bird-catches-the-worm story, replacing the bird with a human and the worm with mushrooms or fish. But there are other stories, like the one I received from Sato Honma from Sapporo, Japan:

> Dear Till,
> "Hayaoki wa san-mon no toku" means "early risers earn 3 mon" (3 mon is about 1 dollar). The origin of this proverb is not clear, but an interesting origin is as follows: Maybe 200–500 years ago, deer were highly treasured in Nara. If a deer was found dead on the property of Nara residents, they had to pay 3 mon. So, the proverb says that if you rise early and you find a dead deer in your garden, then you still have the chance to move it to your next-door neighbor's property.
> Sato

My learned friend Amitabh Joshi from Bangalore surprised me with his statement that such proverbs apparently do not exist in Hindi or Urdu:

> Dear Till,
> Although waking up early has traditionally been considered "good," there is no saying to this regard, to the best of my knowledge, in either Hindi or Urdu.
> There is, however, a couplet from a Punjabi Sufi poet, Baba Bulley Shah, to the effect that if God could be attained by getting up early, surely the roosters would have found God.
> Cheers,
> Amitabh[1]

Early-morning morals make perfect sense in cultures where the entire population shares common rest and activity times and where productivity is predominantly restricted to those hours of the day that are lit up by the sun. In those cultures, the availability of resources has not only spatial aspects (*where* to find the food) but also a temporal aspect (*when* to find the food). Time of day has been of ecological importance ever since different individuals and organisms started to compete with each other. Early-riser morals probably reflect a biological truth that actually formed the evolutionary basis for why the family members in the preceding chapter are (time-)worlds apart.

The Russian version of the early-riser wisdom is an excellent example of how temporal ecology in human populations leads to a temporal economy. "The early riser gathers mushrooms, the sleepy and lazy one goes later for the nettles." Thus, if the early birds gather all of the mushrooms before competitors appear on the scene, they have enough mushrooms to feed their own families and a surplus to sell on the market—a twofold economic advantage. The early riser gets the food for free and can even sell the food to the sleepy and lazy ones—thus, "the morning hour has gold in its mouth."[2] A modern interpretation of the proverb would be "Larks are rich" or "If you want to be rich, be a lark."

The twenty-four-hour day is a circular event and, as such, represents a *temporal structure*. Generally, time has no structures except for those we have created artificially, such as minutes within an hour. Unlike the temporal structure of a twenty-four-hour day, minutes and hours have no concrete beginning or end—they are mere units on a stopwatch or timer. The temporal structure of a twenty-four-hour day is tightly linked to the rotation of the earth and the consequential periodic exposure of its surface to the sun. It has a beginning and an end—the end coinciding with the beginning of the next day. It doesn't really matter where we set the beginning of the day in relation to sun time. Letting the day begin with dawn makes sense since this more or less coincides with the onset of human activity. It has, however, the disadvantage that the time of dawn changes with the season (except for locations on the equator), earlier in summer and later in winter. To avoid this problem one could let the day begin at noon, which is constant throughout the year.[3] But then, midday isn't really the best choice for the beginning of the day. That is why we have chosen its counterpart, midnight, as the zero hour. There are excellent books that describe the history of how humans have dealt with time and how they have organized time, so I will not go into more detail here.[4]

The twenty-four-hour day is not the only time structure on earth—there are three others. The tides—caused by the interactions between moon, earth, and sun—concern all those organisms living on the ocean's coastlines. The tidal peaks and troughs recur every 12.5 hours. The period length of the lunar cycle is 28.5 days, and the duration of a year is 365.25 days.[5] One might think that years were counting days, but that is not the case since the year does not depend on the earth's rotation around itself but on its orbit around the sun. The fact that this orbit takes a quarter-day longer than 365 days is the reason for introducing leap years. Theoretically, our planet's spinning around its own axis could produce different day lengths without affecting the length of a year. As a matter of fact, that has been the case throughout earth's history; many millions of years ago days were shorter than they are today by several hours. The spinning of the earth is slowing down, so our days will be longer than twenty-four hours far in the future.

The temporal structures of tide, day, month, and year deeply affect the earth's environment and therefore also have a strong influence on all forms of life. But not all organisms are exposed to all four temporal structures. Except for some cave-dwelling organisms or those living at great depth in the ocean, practically all beings are exposed to the endless repetition of twenty-four-hour days as well as to the seasons of the year.[6] Tides are only important for organisms that live in tidal zones or for those that feed on tidal organisms. Finally, there are relatively few organisms for which the waxing and waning of the moon's light forms an essential part of their existence.

When organisms are exposed to a regularly changing environment, it is advantageous for them to adapt to these temporal structures, to be prepared for and to anticipate the regular changes. Temporal structures provide the only context in which we can actually predict the future. If I were to bet a large sum of money that a certain series of numbers would be drawn in next week's lottery, you would immediately bet against my prediction. If, however, I bet you that the sun will rise tomorrow, you would just laugh at me. You would certainly not bet against my prediction. The predictive power within temporal structures is an advantage that drove the evolution of biological clocks.

In circular time structures, simple, orderly sequences are difficult to determine—the old chicken-and-egg problem. Is dawn before dusk or is dusk before dawn? Is midnight before noon or is noon before midnight? It seems easier when events lie closer to each other: dawn occurs before noon, dusk after noon. But even these apparent certainties are not necessarily correct. If the cooling-off at the end of the day influenced the weather on the next day, then the rain at lunchtime would be (partly) "caused" by the prior dusk, which would then be before and not after noon—the day is indeed a circular structure. This temporal chicken-and-egg problem is a good reason for challenging the early-bird wisdom. As long as all individuals have similar daily routines (in other words, as long as they are similar chronotypes), the earliest bird has an advantage over anyone getting up later. This was probably true for most preindustrial societies; hence the persistence of the early-bird proverbs. If the distribution of chronotypes becomes as broad as that shown in the first chapter for Central Europe today, then the temporal chicken-and-egg problem starts to apply to the hunt for resources. A small number of very early birds in the population would wake up on their own between four and five in the morning, but even more very late chronotypes would still be awake. There is no reason why these extreme late types couldn't gather all the mushrooms before the early risers arrived in the forest. They could then go to bed and sell the mushrooms to the early birds in the afternoon. They would even have a monopoly on mushrooms because the next good crop wouldn't have grown until the next morning. If you have difficulties with the mushroom metaphor, then think of the impact that stock exchanges have on each other. The last stock exchange of the day, Wall Street, influences the first stock exchange of the next day in Tokyo, which in turn will have an impact on all the others between Tokyo and New York.

This myth that early risers are good people and that late risers are lazy has its reasons and merits in rural societies but becomes questionable in a modern 24/7 society. The old moral is so prevalent, however, that it still dominates our beliefs, even in modern times. The postman doesn't think for a second that the young man might have worked until the early morning hours because he is a night-shift worker or for other reasons. He labels healthy young people who sleep into the day as lazy—as long sleepers. This attitude is reflected in the frequent use of the word-pair *early birds* and *long sleepers* (as mentioned by the journalist).[7] Yet this pair is nothing but apples and oranges, because the opposite of *early* is *late* and the opposite of *long* is *short*. Although duration and timing are the two major qualities of sleep, they are independent from each other. Sleep duration shows a bell-shaped distribution within a population comparable to that of sleep timing, but in this case, more short sleepers are on the left of the distribution than long sleepers on the right.

Almost a quarter of the population sleeps around eight hours (averaged over work and free days); close to 60 percent need between 7.5 and 8.5 hours of sleep (the three most populated categories in the graph). People who get by on less than five hours are very rare (but do exist), as are those who need more than ten hours every night. Due to the different sleep needs of individuals, the concept of mid-sleep was introduced to characterize *when* people sleep, which also gives us an indication of the relationship between an individual's internal time and local (external) time. There are just as many short and long sleepers among early chronotypes as there are among late chronotypes; or turned around, there are just as many early and late chronotypes among the short sleepers as there are among the long sleepers.

Thus, the notion that people who get up late sleep longer than others is simply wrong. This judgment presumes that all people go to

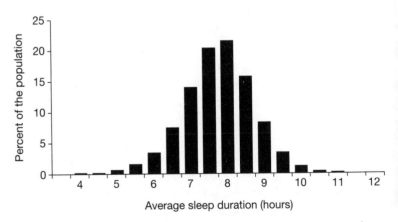

Sleep duration shows a bell-shaped distribution within a population, but there are more short sleepers (on the left) than long sleepers (on the right).

bed at the same time, which we know isn't true—certainly not in the world we live in today. But what is it that makes us fall asleep? Is it merely a signal by our biological clock? Surely not, otherwise we couldn't have an afternoon nap or a siesta. There must be more to falling asleep.

3

Sergeant Simon Stein lay down on one of the mattresses lined up in long rows on the floor of a big windowless gym and wondered why he had signed up for this project. He and thirty-four other soldiers had to perform a lot of psychological and physical tests at different times of the day over a certain period—exactly how long the project would last was still an open question. They were allowed to sleep one-third of the time but were tested during the rest. When first proposed, this sounded almost better than their usual routine, because Stein's unit often had to work shifts that lasted much longer than two-thirds of the day and began at odd, constantly changing hours.

He had finished his first series of tasks and was now allowed to lie down and sleep. A siren gave them just enough time to get comfortable for their rest period before the lights went out in the big gym. Now that Sergeant Stein was lying in the dark on his mattress, however, he couldn't find sleep, and so he reflected on this first testing period. He remained awake, and before he knew it, another siren summoned him and the other volunteers for the next round of tasks, shortly after the overhead lights had gone on again.

The tasks during the light period were not too difficult. Some of them were actually quite a lot of fun. There were reaction-time tests where they had to press buttons, a left or a right one, to match two little lights mounted on a vertical board in front of them. They had to estimate time spans, in the second and minute ranges, or they had to use their full muscular strength pressing a flexible barbell. They had to cross out all the p's on a sheet filled with p's, q's, and d's. They had to do simple arithmetic, memorize shopping lists, or ride a stationary bike as fast as they could. The test periods were crammed with so many tasks that Sergeant Stein usually felt as if they were over before they had started.

The alternation between light and darkness, between rest and activity, had repeated itself for several cycles now, and since Stein had hardly ever managed to fall asleep during his rest periods—no matter how many sheep he counted—he began to feel pretty tired. Finally, a couple of cycles later, he fell asleep almost before his head hit the pillow and slept like a log until the siren and the lights made him sit upright with a jerk.

Although he was quite sure that he had fulfilled all of his tasks as well as he had done at first, he somehow got the feeling that his supervisors were not as enthusiastic as they had been. Officially the volunteers didn't receive any feedback from the supervisors, but Stein was an observant man and he was sure that the attitude of his supervisors had changed. Maybe they were also worn down, although they frequently spelled each other, unlike the test subjects.

The longer the project lasted, the more tired he became. It was astonishing that he was unable to get a wink of sleep during some cycles but was able to sleep the entire rest period during others. He lost track of how many days they had been alternating between performing tasks and more or less successfully getting some sleep between the testing cycles. But he was starting to see a pattern in his changing capacity to fall asleep. He was puzzled by the fact that, despite this insight, he was unable to sleep during some cycles—because he was more worn out than he had ever been in his life. His mates were all experiencing similar difficulties and patterns of exhaustion. Finally, the project was terminated because the volunteers were deeply exhausted.

Sergeant Stein's story represents many different experiments that investigate how different qualities of an individual change in the course of the day. In these types of experiments people live under very controlled conditions for several days and perform countless tests that probe performance, reaction time, and muscle strength across the twenty-four-hour day. Stein's experiences are closely related to an experiment performed by the Israeli military. The military powers of all nations are extremely interested in understanding what makes humans sleepy and, even more important, why lack of sleep affects us so profoundly and so negatively. In combat situations, not needing to sleep or taking a drug that prevents sleep-loss from interfering with performance would provide an obvious advantage.[1]

Chronobiologists, those researchers who investigate biological clocks, are mainly interested in sleep because its timing is the most conspicuous expression of the body clock in humans and other species. There are many other fascinating aspects of sleep besides its timing, such as sleep structures, the function of sleep, sleep pathologies, or the relationship between sleep and the immune system, to name just a few. Many of these aspects are scrutinized by sleep researchers. Chronobiology and sleep research used to be entirely separate disciplines, but their representatives came to recognize that both groups can learn and profit from one another, and nowadays they often meet at the same conferences. An excellent example of teamwork that brought the two disciplines much closer together is the theory about how sleep is regulated. It was developed by the sleep researcher Alex Borbely and the clock researcher Serge Daan.

Both knew that sleep is regulated by at least two independent components—by being tired and by time of day. The first component was known long before people ever heard about biological clocks. The longer we are awake, the more tired we get—it's common sense. But even before the discovery of the biological clock, people must have found themselves lying awake despite utter exhaustion (like Sergeant Stein). Borbely and Daan consolidated the two reasons why we fall asleep into one relatively simple model. This model involves two different kinds of rhythm produced by two different kinds of oscillators, one behaving like an hourglass and the other like a pendulum.[2] While an hourglass has to be actively turned once its top chamber is empty, a pendulum swings by itself (forever, if it is kept in a vacuum).

In Borbely and Daan's model, one of the chambers of the hourglass represents sleep pressure. During wakefulness, this chamber is at the bottom and is continuously filled with "sand." When the level of sand reaches a critical threshold, we fall asleep. The hourglass is turned around, and the chamber empties until it reaches another (lower) threshold that wakes us up; the hourglass is turned around again, and the cycle restarts. These alternating processes create a sawtooth pattern.

So far so good, but this model doesn't explain why we can have a long sleep at some times of day, even if we are not exhausted, whereas at other times, we can only get a short nap or are even incapable of

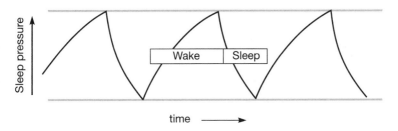

The sawtooth pattern of the sleep–wake cycle.

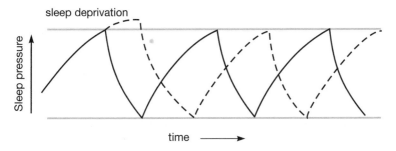

Staying up just a little too late—hypothetically.

falling asleep, despite being totally exhausted. The model also cannot explain why we always sleep more or less at the same time of the day (or rather night). If we stayed up beyond the upper threshold (depriving ourselves of sleep) just once, we would—according to this model—go to bed a little later in the evening for the rest of our lives.

This doesn't fit with our experience. Despite wiggles and larger perturbations in our bedtimes, most of us tend to fall asleep within a relatively stable time frame, with the majority of the population sleeping while our part of the globe is turned away from the sun. Daan and Borbely solved this problem by postulating that the thresholds that switch our body into sleep and back to wake oscillate with a daily rhythm.[3]

This model now explains why sleep deprivation does not permanently shift our sleep timing: once we get recovery sleep, the rhythmic shapes of the thresholds will always move our sleep back to its usual times. Thus this model is much closer to what we experience in real life.

Now that you have an idea of how scientists think about sleep regulation, we can turn back to Sergeant Stein, who was allowed to sleep one-third of every cycle. Why did he get so tired? Don't we all, on average, spend about one-third of our sleep–wake cycles asleep? That alone cannot be the reason he and his fellow soldiers became so exhausted that the scientists decided to terminate the experiment. Were the test batteries during the wake periods too exhausting? It's unlikely; after all, the participants were all highly trained, fit soldiers. The reason why they became so exhausted was that they couldn't get enough sleep over the course of many cycles. Their experimental "days" (or cycles, as I call them) were only thirty minutes long. They worked on tests for twenty minutes, then a siren signaled them to go as fast as they could to their mattresses, then the lights went out and they were allowed to sleep for ten minutes. If sleep were regulated purely by the average time a person was awake, the soldiers should have had nothing to complain about because, on average, the proportion of sleep to wake matched the usual proportion outside the study.

But sleep regulation is more complex: they didn't sleep during the first hours of the experiment because their sleep pressure hadn't reached the upper threshold. Further into the experiment their sleep pressure reached the point that triggered sleep, but instead of sleeping for eight hours, they were woken up after ten minutes for another twenty-minute round of tests. What the model of Daan and Borbely doesn't show is that there are times during our internal twenty-four-hour day when we cannot readily fall asleep, despite being utterly

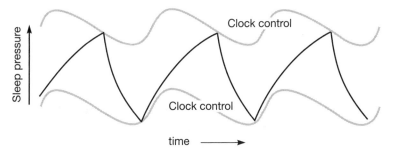

The thresholds that switch our body into sleep and back to wake oscillate with a daily rhythm.

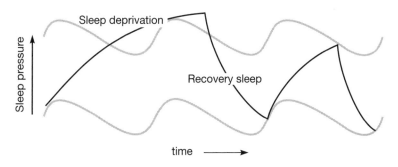

This rhythmic threshold helps our sleep to stay synchronized at its normal phase, even when we stay up too late.

exhausted. These times usually occur during our daytime, and those were the times when Sergeant Stein couldn't find sleep during the allotted ten minutes. He might have fallen asleep later, but that was prevented by the experimental plan. In the end, the soldiers were so sleep-deprived that the experiment was terminated.

Interestingly, there is one time during our daily cycle when we are least likely to fall asleep, and that is shortly before our body clock opens the temporal window that allows us to sleep best. It almost seems that this time is a forbidden zone for sleeping.[4] This zone was first discovered by the experiment described in our case story.

Sleep deprivation has often been used as a form of torture, to make prisoners confess their supposed crimes. In most cases that torture works really well. The only drawback is that prisoners get so exhausted that they practically lose their minds and start to confess to things they never did. I once received a phone call from a police interrogator who wanted expert advice about how and when he could use sleep and time of day to weaken the defense mechanisms of a suspect. I told him (in somewhat politer words) to go to hell or, even better, to call Amnesty International.

This chapter has covered the different factors regulating sleep. We don't fall asleep just because we are exhausted; there is another factor that controls our capacity to fall asleep and to benefit the most from our sleep, and that factor is the biological clock. But what is this biological clock? How does it work, and how important is it for us in our daily routines? And who was the first scientist to discover this extremely important part of our biology?

A Curious Astronomer

4

On a lovely evening toward the end of the summer of 1729, the French astronomer Jean Jacques d'Ortous de Mairan sat at his desk working on a manuscript. He paused to think about a difficult sentence. As some of us do when we concentrate and turn our thoughts inward, he gazed out of the window into the cloudless evening sky. Although it was still bright enough to work without a lamp, the quality of the celestial light clearly announced the approaching night. He didn't manage to capture his thoughts in words that his readers would unmistakably understand. His inward gaze therefore slowly came around to focus on the world in front of his eyes, which fell upon a plant with delicately feathered leaves growing in a pot on his windowsill. The mimosa was one of his favorites. Now that his attention had returned to reality, de Mairan saw that the mimosa had already "gone to bed" by furling its leaflets, which had been stretched out throughout the day. He envied the plant for its capacity to fall asleep with unbreakable regularity and precision. He wondered whether one could compare the up and down movements of plant leaves with wake and sleep in humans—which unfortunately did not show the same regularity for him. Particularly when he was writing, he worked into the small hours, until he was too tired to hold a quill. Even then he would lie awake for a long time with his thoughts whirling around the manuscript until he finally fell asleep, waking only well after the sun had risen. This particular evening, while trying to find sleep, his mind got hooked on the question of whether sleep and wakefulness had any similarity with the daily up and down of his mimosa's leaves. He was sure that most animals slept, so why not plants? Yet plants didn't run around, and they didn't exhaust themselves like animals did, so why should they recuperate in sleep? Being sessile, they couldn't escape the sunlight like animals. They were completely dependent on light and darkness. Maybe it was strenuous for the plants to keep their leaves in a horizontal position. Did they expand them during the day to catch the sunlight or to shade their lower parts? In either case, it was unnecessary to keep them up once the sun had set. The leaf movements evidently had something to do with light and darkness. But why were the leaves then already folded down before the sun had set? Obviously the direction of the sun changed over the course of the day (for example, the sunlight didn't come directly from above in the early evening hours), so it was advantageous to adapt the leaves' position, whether to catch light or to create shade. He had often observed how plants turned their leaves toward the sun, tracing its path across the sky. The more he thought about the phenomenon of leaf movement, the clearer it became to him that it all depended on sunlight and darkness.

Suddenly he was wide awake, even though it was 3 A.M., and he sat up quickly. He had an idea about how to determine if the leaf movements were merely passive reactions to the sun. He got out of bed and ran downstairs to his study. Frantically, he opened the door of his desk cupboard and pulled out all the drawers to make space for the mimosa, placed the pot inside, and closed the door. He was so excited about his simple experiment that he feared he would not sleep at all that night, but once back in bed he fell asleep immediately. It was well into the next day when he woke. It took him quite some time to remember that he had started an experiment the night before. He jumped out of bed, went to his study, and closed the curtains to make the room as dark as possible. Then he cautiously opened the desk door just wide enough to get a glimpse inside. To his great surprise, the mimosa was "awake" and had fully extended its leaflets to their usual daytime position, even though they had been in complete darkness!

De Mairan had a hard time concentrating on his manuscript that day. It was again beautiful, and he decided to work outside so that he could keep the curtains in his study drawn. Every hour or so, he went inside and peeked into the darkness of his desk. The leaflets stayed in a horizontal position throughout the entire day, started to fold downward in the late afternoon, and were fully collapsed even before the sun had completely set. When he went to bed at around midnight, the mimosa was fully "asleep."

Over the course of the next few days, de Mairan remained so excited by his observations that he slept very badly. After almost a week of severe sleep deprivation, he was so tired that he fell into a deep sleep sometime between 9 and 10 P.M. He woke up once during the night, looked out of his bedroom window, and saw that, though it was still dark, dawn was beginning to break. He went downstairs to have another peek into his desk. His study, with drawn curtains, was still pitch black and he had to light a candle to find his way. He put it down as far away from the desk as possible. Being adapted to the night's darkness, he was able to see very well in the barely lit room. He opened his desk's door a tiny crack, enough to see that the mimosa had already started to lift its leaves. He continued his little experiment for several days and found that the mimosa's leaf movements continued with the same regularity as if it were still standing on the windowsill.

The plant evidently continued to "know" the sun's position and continued to be "aware" of night and day. During the later stages of his experiment, de Mairan tried to keep an early schedule so he would be able to open the door not only after the sun had set but also before any sunlight could enter his study through the tiny slits between the curtains. Every morning before the sun had properly risen, the leaves were up, and every evening just before the sun had set, they were furled. The mimosa was like a sick person who lies in bed in semidarkness for days but still sleeps during the night and is more or less awake during the day without ever seeing true daylight. He concluded that leaf movements, at least those of the mimosa, were not merely a reaction to light and darkness.

Throughout the remainder of the summer of 1729, de Mairan repeated his experiment many times. He allowed the plant to vegetate in constant darkness for only a couple of days at a time before he exposed it to the sun again, knowing its need for light. He even borrowed mimosa plants from friends to repeat the same experiment and always found that their leaves' movements continued in constant darkness. He became so fascinated by his findings that he abandoned the manuscript he had sworn to finish by the end of the summer and, instead, wrote down his "botanical observations." It would be interesting, he thought, to see whether other plants behaved the same way. It obviously wasn't light that controlled the up and down of the leaflets—maybe it was temperature, which could easily be tested by using an oven. Even though the rhythm of leaves apparently came from within the plant, independent from the outside light, he wondered: could one still reverse the leaves' rhythm by reversing the natural course of day and night with the help of artificial light?

De Mairan's botanical observations are the first traceable scientific source that mentions the independence of daily rhythms from the changes of night and day, of dark and light. His published accounts constitute one of the most concise scientific papers, consisting of fewer than 350 words and, as typical for the French writing style of the time (or even of now), of only seven rather long sentences. His observations were reported to the French Royal Academy of Sciences in the same year he performed the mimosa experiments by his distinguished colleague M. Marchant, a fellow of the Academy, and were subsequently published in the Academy's proceedings. The paper's last two sentences state that the daily occupations of M. Mairan (namely, astronomy) had prevented him from conducting any of the experiments proposed in the paper. He contents himself with a simple invitation to botanists and physicists to do so, even though they themselves may have other assignments to complete. But he warns them that the progress of true science is inherently slow.

Everything in de Mairan's paper turned out to be true, including his last statement about the speed of scientific progress. Although de Mairan's observations were sporadically picked up by other botanists and zoologists (including Darwin), it took 200 years before botanists, at first, then zoologists, and then human physiologists started to investigate with modern techniques the mechanisms behind the enigmatic internal clock discovered by a French astronomer who was far too busy to pursue his observations with more experiments.[1]

5

The young man sat at his desk looking extremely pleased with himself. He was on schedule; in two weeks his thesis would be finished. He got up and walked through his windowless studio apartment to the small kitchenette to put on some water for a cup of coffee. While waiting for it to boil, he added a couple of items to a shopping list lying on the kitchen counter—coffee, milk, butter, . . . After checking his list once more, he opened the solid, sound-proof door of his apartment and placed the piece of paper on one of the shelves in a small, dark walk-through cupboard outside of his apartment. The cupboard led to another door, now closed.

On his way back into the apartment, which had served him as bedroom, study, dining room, and living room for many weeks, his expression changed—he had remembered something. He walked quickly to his desk where he pressed a button, as if he were ringing a bell for room service. Nothing happened, and about one minute later he pressed the button again. Throughout the day he had repeated this procedure fifteen times. Although nothing ever happened after pressing the button, he did not appear to be surprised. With a sharp whistle, the kettle begged to be taken off the stove. The young man turned swiftly and almost stumbled over the long cable leading from his belt to a plug in the wall. While he was pressing the button he had changed his mind about the coffee—he would instead make a bowl of instant soup. He had been writing all day and was tired.

Once he had eaten his supper, he undressed, took a shower in his tiny bathroom, and climbed into bed, reaching for his diary on the bedside table. Conscientiously and meticulously he described all his activities, his physical state, and his feelings over the past sixteen hours. Then he got out of bed and pressed the button on his desk once more and, when he thought a minute had passed, one last time for today. He slipped between his sheets and buried his head in the pillows, without turning off the overhead lights covering most of the ceiling behind a layer of milky glass.

———————

While the young man was sleeping, a young woman carrying an empty tray went down a flight of steps leading into a small hill. It was a brilliant spring morning, and she had just arrived at work. After opening a thick and heavy door she entered a dark room filled to the ceiling with scientific equipment and cables. She switched on the lights, looked at the large plan of two apartments mounted to one side of the wall, and checked the different lamps and settings. Having entered all the collected information in a notebook, she turned and opened another door giving way to a dark walk-through cupboard lined with shelves. She retrieved some capped bottles and a small piece of paper. Placing everything on a tray, she walked out of the hill, closing all the doors behind her, and proceeded to a larger building. Once inside, she entered one of the rooms leading off a long corridor lined with wooden, built-in cupboards and placed the tray on the bench top filled with laboratory equipment.

At the other end of the laboratory, two scientists were discussing data, and one of them turned to the young woman who had just entered the room.

"How's everything down in the bunker?"

"He's asleep, but his list says he needs a few groceries. As soon as I've analyzed the urine samples, I'll pop down to the shops."

"No need for that, Karen, it's his last day. In about ten hours, once he wakes up, we'll go in. He'll be in for a surprise."

"So I'd better go and get today's newspaper," Karen replied with a broad smile.

———————

Several hours later, the young man woke up. He felt refreshed, and cherished the fact that, like every day during these long weeks, no alarm had woken him. Like yesterday evening, he got out of bed, went over to his desk, and pressed the button once and then again a minute later. To his great surprise, for the first time, his action prompted a response—the door opened and three scientists entered, an elderly professor and two coworkers.

"What are you doing? Has anything gone wrong? Why are you terminating the experiment? Didn't you say there would be no contact until the two months were over, or if I decided to quit?"

"Yes, we did," said the elderly professor, "and we aren't breaking the rules. Congratulations and sincere thanks for a very successful experiment—your lonely weeks are over."

As if on cue, the young woman who had previously collected the urine samples produced a newspaper with a broad smile and handed it to the astonished young man, who looked at the front page and read the date.

"This is a practical joke—there's no way I can be wrong by two weeks. I knew that living without a clock would probably get me out of synch with the outside world, but . . . What time is it?"

He asked this last question with great eagerness, realizing how little he had missed the usual temporal certainty during the past weeks, but how important it had suddenly become as soon as he noticed the wristwatches everyone was wearing.

"It's 8 P.M., the fourth of April. You've lived in the bunker for sixty-three days as one of our best subjects. While all of our days had exactly twenty-four hours, many of your days were as long as forty or even fifty hours."

"How can that be? I pressed this button twice every time I thought an hour had passed, and I never did that more than sixteen times a day. Does that mean I wasted most of my time sleeping? I thought I was keeping to such a strict schedule on my thesis," he lamented.

"No, as a matter of fact, you never slept more than a third of your days, just like in real life. When you were awake sixteen hours, you slept eight hours; and when you were active for thirty-two hours, you slept sixteen; but during those long days, the time span you thought to be an hour doubled, and so you never noticed the change."

"But I never had more than three meals per day. I never felt especially hungry, and I never ate any extra big meals."

"That is why we are so intrigued with this experiment. When you extended your days you truly expanded time: your hour estimation doubled, and you kept to your habitual meal times although that meant not eating anything for as long as sixteen hours. Despite this, according to our records, you haven't lost weight. We don't know yet how to explain all this, but the contacts on your apartment floor tell us that you were less active during those long days; maybe that begins to explain why you didn't lose weight despite eating about half the amount of calories per twenty-four hours. The only function that didn't go along with the extended days of wake and sleep was your body temperature. It continued through your long days with its usual twenty-five-hour day. So, while your sleep–wake cycle occasionally lived through forty- to fifty-hour days, other parts of your body kept to a circa-twenty-four-hour day."

The young man forgot his initial shock at having lost almost two weeks of his life and started to listen to the scientist's description with growing interest.

"During the first week of your experiment, when you still had a watch and when we kept the doors to the bunker open, you stayed synchronized to the normal twenty-four-hour day. You went to bed at around midnight, slept through your temperature minimum at approximately 4 A.M., and woke up after about eight hours of sleep. Then, we took all of your clocks away except for the one in your body, and you started to live your own days—your body clock began to 'free-run.' You gradually stayed up longer, approximately 1.5 hours every cycle. Your temperature rhythm also became longer, but only by about an hour. After a couple of cycles, you went to bed around the time when your temperature rhythm hit a trough.

"For the next week, your activity–rest cycle and your temperature rhythm remained synchronized with each other, both producing days of approximately twenty-five hours, and you always went to sleep when your temperature was at its lowest point. This is what most subjects in the bunker do, but in your case, after about ten days into your clockless existence, your activity–rest cycle started to run at a slower pace than your body's temperature rhythm—the two rhythms uncoupled. For many days, they appeared to take no notice of each other. Only when your sleep–wake rhythm was delayed so much that the time you went to bed came near your temperature minimum again did the two rhythms resynchronize for a couple of cycles before they broke loose again."

The young subject had millions of questions, but the professor suggested they walk up the mountain to the beer hall of the monastery to celebrate the end of a great experiment, and there they could answer his questions—if, he said, winking, they knew the answers.

Botanists had returned seriously to the century-old question of how internal clocks work in the 1930s. They had proven in many experiments that de Mairan's observations were real and had worked out many rules about the behavior of this biological clock under different light–dark or temperature cycles, predominantly in plants. However, biological clock research only received scientists' full attention after the Second World War. The two main pioneers of the young research discipline were Jürgen Aschoff in Germany and Colin Pittendrigh, who worked at Princeton University at the time and later joined Stanford University. In the early 1960s, Aschoff was appointed one of the directors at one of the research institutes of the German Max Planck Society initiated by the founders of modern behavioral physiology, Erich von Holst and Konrad Lorenz. The newly created institute was situated in the heart of Bavaria near the "holy mountain," one of the best-known traditional beer-brewing monasteries, Kloster Andechs.

Aschoff and his colleague Rüdger Wever wanted to investigate whether humans were also governed by an internal timing system, by a body clock, as had been shown for many different plants and animals. Although most of the scientists who investigated biological clocks had no doubt about their existence, some researchers still believed that one could never create a completely time-free environment. They argued that the observed ongoing rhythms could still be controlled by some unknown factor linked to the rotation of the earth. Aschoff and Wever, therefore, set out to build the Andechs "bunker,'" two small time-free apartments inside a hill. The interior of the bunker was shielded against everything that could disclose any time-of-day information to the subjects. It had no windows, and was completely soundproof and shielded against vibrations caused by the heaviest vehicles driving on nearby roads. It was even equipped with a metal cage keeping out the more-or-less regular changes of the earth's electromagnetic field.

The apartments could be entered only through a corridor separated by two thick doors, each of which could be opened only if the other one were closed. This hallway served as a link between the time-free world inside to the time-driven world outside. Subjects placed their shopping lists or bottles with urine samples on the shelves of this corridor. The urine was used to monitor the daily rhythms of metabolites, such as potassium or calcium, as well as substances that allowed the scientists to estimate ups and downs of hormones. The scientific staff looking after the bunker experiments and their subjects entered the corridor with high irregularity, even at odd hours during the night, so that the subjects could never deduce normal work hours by the times when their shopping lists and samples were collected, or when the requested items and new empty sample bottles were delivered. The floors of the apartments contained electrical contacts that were used to record when, and how much, the subjects moved. To record their body temperature, subjects had to wear a rectal probe, which—in that pretelemetric era—was connected by a long cable fixed on a belt around their waist to a wall socket. Subjects were asked to keep detailed diaries about how they felt both physically and psychologically.

Depending on the scientific question, the duration of the experiments varied between one and several weeks. Sometimes the apartments were separated into two completely independent units, each inhabited by a single subject; other experiments involved a group of people and investigated how the body clocks of different individuals influenced one another.

Although free-running daily rhythms had been described extensively in plants and animals by the time the bunker experiments started, the results from putting humans in temporal isolation were almost eerie—even for those scientists who had hoped that the biological clock in humans would behave like that in other creatures. During the first days of each experiment, subjects remained in contact with the normal daily routine of the outside world, with regard to both staff and daylight. Once they were isolated from any time cues, their daily routine continued almost normally with its usual structure: two-thirds awake and one-third asleep. One of the important differences between the real world and life in the bunker was that the periodicity of the self-chosen bunker days wasn't exactly twenty-four hours but in most subjects slightly longer. Another important difference was that subjects usually went to sleep at around the time when their body temperature hit a daily low point. This varies from what we do when we live in the real, time-driven world: our body temperature hits its low point about halfway into our night's sleep.

In the majority of subjects who lived for several weeks in the bunker, all bodily rhythms that were recorded oscillated in synchrony.[1] In other words, all had the same period of circa twenty-four hours. In less than a third of the subjects, however, an unforeseen picture emerged: different rhythms oscillated with different periods and, therefore, drifted away from each other. While the minimum core body temperature recurred about every twenty-five hours, the subjects went to bed every forty hours. It seemed as if different outputs of the internal timing system had desynchronized from each other. This observation suggested that more than one clock controls our daily behavior and physiology.

In most of the subjects who showed this *internal desynchronization,* the sleep–wake rhythm together with the rest–activity rhythm continued at a slower pace than other, more basic bodily functions, such as the waxing and waning of body temperature or of hormones. In a small number of subjects, the desynchronization between the behavioral and the physiological rhythms was the other way around. They lived shorter days in their behavior compared with their physiological rhythms, which in all cases were never far from twenty-four hours. Because the length of the body clock's day was only close to but not exactly that of our normal twenty-four-hour day, the rhythms produced by the body clock were called circadian rhythms.[2] This rather scientific-sounding term even made it into the pop charts in 1998 in the song "Daysleeper" by the group REM.[3]

Aschoff and his colleagues discovered the internal desynchronization of different body rhythms in the late 1960s. The activity rhythm in animals can adopt quite long or short periods in temporal isolation. These can even split into two distinct components, each running at its own period, but a clear separation between the rhythms in behavior and physiology seems to be a human specialty. Some subjects in the bunker lived through days twice as long as our normal days and in some rare cases even longer. Yet in all cases, the period of the core body temperature rhythm and other physiological rhythms stayed close to twenty-four hours.

Aschoff's hypothesis of internal desynchronization was often criticized, based on the following argument. In the case of an apparent internal desynchrony, the period of the sleep–wake cycle is often double that of the body temperature rhythm. Consequently, subjects experience two temperature minima during each of their behavioral days, one during their nocturnal sleep episode and another in the middle of their active period. Many of these subjects took extended afternoon naps during their extra long bunker days approximately around the time when their body temperature hit a second, "daytime" low. The critics argued that the nap was in reality a true nocturnal sleep episode, misinterpreted by Aschoff and his team.

Although the "misinterpreted naps" argument is valid and puts the human clock in line with that of other animals, there are several observations that support Aschoff's hypothesis of internal desynchrony. As described at the beginning of this chapter, the subjects who experienced these extra-long days never mentioned anything unusual in their bunker diaries. They ate only three main meals throughout these long days and didn't double their portions (which one would expect after a doubled day); in addition, they went on average to the toilet to defecate only once during their extra-long days. The most compelling result supporting the concept of internal desynchronization concerned the passing of *subjective time*. To monitor how the time-free environment affected subjective time, subjects had to perform time estimation experiments.[4] Subjects in temporal isolation behave quite similarly to other isolated individuals whom we all have read about, for example Daniel Defoe's Robinson Crusoe. They become extremely observant and try to keep track of the time of day and especially of the exact number of their isolation days. So bunker subjects were ideal in that they recorded anything out of the ordinary. If, for example, subjects realized that they had pressed the hour-button not the usual sixteen times per day but as often as thirty-two times, they would have noted this in their diaries. Indeed, Aschoff's recordings showed that the estimation of one hour went along with the length of the days they lived in the bunker, even if they lasted as long as fifty hours. Their short-interval minute estimation, however, didn't change along with their days' length. The stability of the short-interval time estimation offers yet another explanation why subjects had no clue as to the actual length of their bunker days. They would have immediately noticed if their short-time perception had changed. For example, the music they were allowed to listen to should have sounded strange. ("I have the feeling that my record player is on its way out, because it runs fast" could have been an entry in their diaries.)

Despite clock researchers interpreting the results of the Andechs bunker differently, these experiments have introduced us to a fascinating world of internal time. They showed that each individual's body clock responds quite differently to the time-free environment. Every subject adopted his or her very own individual internal day length, and some even lived by more than one internal time frame, with their physiology "listening" to one and their behavior to another. Aschoff's notes showed that subjects felt best if bodily functions were in synchrony.

You may have asked yourself why evolution was so sloppy as to create a biological clock that cannot keep track of time properly. Even without internal desynchrony, the internal days produced by the body clock of bunker subjects rarely lasted exactly twenty-four hours. The biological clocks of most plants and animals show similar deviations from the twenty-four-hour day when kept in temporal isolation. Evolution has produced impeccable brain functions, which enable us to learn to hit the center of a dartboard from a great distance. It has selected for fine motor skills, so that we are capable of keeping an exact tempo on a musical instrument. And it has equipped us with auditory skills that can distinguish two sound clicks separated by milliseconds. So, why hasn't it created a more exact body clock?

One could argue that exactness is apparently not necessary for a good body clock, otherwise it would surely have been the outcome of evolution. Yet this argument is circular since it postulates a perfect evolution. There is actually a much simpler and more logical answer to the question of clock evolution. Ever since de Mairan discovered the ability of biological clocks to run free in constant darkness, researchers have been fascinated by the fact that the daily rhythms are not merely a reaction to night and day but are obviously generated by an internal mechanism. Researchers remain fascinated with biological rhythms continuing unabated in constant conditions, and they still investigate many properties of the body clock in time-free experiments. Yet, organisms never encountered a time-free environment over the course of evolution. With few exceptions, such as cave dwellers or creatures on the bottom of the ocean, organisms have always experienced brighter, warmer days and darker, colder nights. Even these persistent daily changes can vary a great deal over the course of a year in most parts of the world. The biological clock did not evolve in a time-free world, and constant conditions were never part of the selection pressure driving the evolution of the body clock. Thus, evolution's rules of chance and necessity could not act on the development of a biological clock that continues with a precise twenty-four-hour rhythm in constant conditions.[5] But why do biological clocks continue to oscillate without time-of-day information in the first place? This ability merely reflects that the way clocks evolved under an endless stream of daily changes has formed them in such a way that they continue even in a time-free environment.

If I have succeeded in convincing you that time-free conditions show us only how the clock evolved to serve the organism in the real, time-driven world, you may conclude that the bunker experiments described in this chapter are all extremely artificial and have little to do with everyday life. Yet, there are four important points to take away from these experiments. First, our body's internal day is controlled by its own biological clock (as in most other creatures). Second, because clocks do not generate an exact twenty-four-hour day, they obviously must be periodically set—how else could they be of any advantage in real life? Third, this internal timing system can differ from individual to individual. Finally, we feel best if all of our bodily functions oscillate in synchrony.

6

It was seven o'clock when the alarm abruptly woke Harriet from deep sleep—she must have drifted off only a couple of hours ago. She angrily pressed the snooze button on her alarm clock and turned around again. She fell asleep immediately but was awakened a couple of minutes later by the alarm's insistent reminder. This procedure continued for about another half an hour. Although she sensed that it was once again a magnificent summer morning, promising to be a wonderfully warm day, she felt like an ice cube and dog-tired. In her half-conscious state she wondered how appropriate the expression was because her golden retriever, Sally, was obviously already wide awake, giving her cheek a sloppy kiss. When Sally's morning greetings got too persistent, making it impossible for her to doze off again, she turned off the snooze function, got out of bed, and fumbled her way into the kitchen to get the coffee going while she took a long shower.

Once again, she was living through one of those stages where she couldn't sleep at night, couldn't get up in the mornings, and only fully woke up during the afternoon. These stages repeated themselves relentlessly in a monthly rhythm. For approximately two weeks, the symptoms got worse and then gradually improved again for the rest of the month. When she sat at her breakfast table clutching her coffee mug and unable to eat even half a piece of toast, half-heartedly listening to the radio's morning program, she longed for those one-and-a-half weeks in every month where she slept at night and was fresh during the day. She lost track of time, almost falling asleep again while sitting upright at the kitchen table, when she realized that the 9 o'clock news had already started. After she had gulped down the rest of her coffee—what would she do without caffeine?—Harriet got her bag, put Sally's harness on, and walked out into the staircase of her apartment building. It took her much longer than on her good days to lock the door with her key. Then she took Sally—or rather Sally took her—down the stairs onto the street. She made her way to the bus stop, and when she heard the bus coming she made a cautious step toward the curb, waited until the doors opened, responded to the bus driver's cheerful "Good morning, Harriet and Sally," and took her usual seat. Exhausted, she asked herself why she was so different from those who were otherwise so similar to her.

You may have read that last sentence several times, finding it a bit awkward. The strangeness of this sentence is intentional and holds a clue to the problem discussed here. You may have your own hypothesis to explain Harriet's problems. As a woman, she goes through monthly hormonal cycles, which could be the cause of her sleeping difficulties, but in this case your hypothesis is incorrect—although it could be an entirely probable solution. Harriet's sleeping problems are related to her internal timing system.

As you read in the preceding chapter, the body clock of most humans generates days slightly longer than twenty-four hours. For the sake of simplicity, let us assume that the internal day of a bunker subject is exactly twenty-five hours and that his behavioral and physiological rhythms do not desynchronize. Imagine that he lives his entire life in the Andechs bunker but that he has to work for a living as an employee with a local company via the internet. Further, his hours of work are not different from those of other people living outside the bunker. I have described the many precautions that Aschoff and Wever considered when they built the bunker. These precautions were taken because they just didn't know at the time which environmental factors could influence the body clock, and hence jeopardize the experiments by synchronizing it to twenty-four hours. Now almost fifty years later, we know that we can make the body clock run free by simply keeping subjects in a windowless room, so they do not perceive natural night and day or read the local time from any device. De Mairan, we recall, compared his mimosa plant to sick people lying in bed for days but still keeping to a normal sleep–wake rhythm without ever seeing true daylight. Although the French astronomer discovered the persistence of daily rhythms in constant darkness, he didn't quantify the rhythms of leaf movements and thus missed the fact that free-running rhythms can be longer or shorter than the twenty-four-hour day outside.

The imaginary subject in his lifelong bunker situation has a problem. Since his body clock's day is twenty-five hours long while his work days repeat every twenty-four hours, he has to get up every day an (internal) hour earlier to be punctual for his computer appointments. After twelve days his body time is twelve hours later than local time, so that he has to be at work in the middle of his internal night—he has gradually become a night-shift worker. Another twelve days later, his body time will be again in synch with local time, and he can easily wake up to go perkily about his computer work.[1]

Of course, you immediately recognize the similarity of the symptoms between the subject in the bunker and Harriet. But why is Harriet's life affected as if she is living in the Andechs bunker? The answer is simple: because her body clock is not synchronized to the social world. But why? You may have had another hypothesis, namely that Harriet is blind and that Sally is her guide dog. If so, your hypothesis is absolutely correct. In the last sentence, Harriet asks herself why she is so different from the many other blind people who sleep as well as she does during her best periods every day of the month.

Light is the main signal that resets the body clocks of plants and animals, including humans, to the twenty-four hours of the earth's rotation. Because Harriet is blind, her body clock is not informed about night and day and, therefore, runs free, as if she were permanently living in the Andechs bunker. So far so good; but how are we to answer her question?

Until recently, ophthalmologists were convinced that the eye of mammals was one of the best-understood organs in the history of anatomy and neuroscience. Eyes provide our brain with information about the structure of the outside world and, thus, help us and other animals to find food; to recognize enemies, parents, or partners; even to read these lines. Thus, our entire concept of the eyes' function is focused on vision. Light enters the eye through the lens and, as in a camera, projects an upside-down image of the world onto the back of the eyeball. This part of the eye is covered with a layer of cells, called the *retina,* which consists of millions of tiny light receptors, sampling the picture. In our electronic age, we could say the picture is translated into pixels. Several of these receptors, after communicating with their neighbors (to increase contrast, for example), assemble the collected information into packages that represent the light quality of a given point on the retina and, thus, of a given area of the outside world. These information packages are sent into the brain via the optic nerve.

We have two eyes to provide our brain with enough information to see the world in three dimensions. The two images encoded by the left and the right eye have to be put together in the correct way so that higher regions of the brain can make sense of the puzzle pieces. Part of the optic nerve, therefore, has to cross into the other half of the brain. This nerve crossing lies a couple of centimeters behind the bridge of our nose, forming the *optic chiasm.*[2]

The retina houses two kinds of light receptors, which have been given names according to their shapes. The *rods* are responsible for encoding pictures under dim light conditions and produce a mental image in grayscale. The *cones* are responsible for helping the brain to compute color when more light is available. Light receptors are highly specialized; each type can only detect light within a relatively narrow range of wavelengths (light color). If the retina had only one type of receptor, the brain would be incapable of computing colors —it could only create mental images consisting of brighter and darker "pixels." The reason why we cannot see color under dim light conditions is that our retinas have only one type of rods. The brain can only recognize different colors if it can compare the information coming from at least two different receptor types, each of which are specialized for a different range of wavelengths but "look" at the same spot in the picture. If one of them is specialized on red light and another on blue, and if we were looking at a strawberry, the red receptor would transmit the information "it's very bright," and the blue receptor would convey "it's pretty dark." If we were looking at a blueberry, they would provide the opposite information. Color detection works even better with more than two receptor types. Humans have three different cone types, specialized for red, blue, and green, respectively. The reason why some people are color blind is because one or more of their genes, containing the information for building an important component of a cone type, is defective. I am going into the science of vision in detail because light reception is extremely important for understanding internal time.

All this information and much more had been worked out in great detail by the 1980s, and scientists began to understand how vision works—from the first stages of translating light information into nerve signals (in the eyes) right up to the computation of this information into mental pictures (in the cortex).[3] The visioncentric concept of the eye received a big shock when Russell Foster, a British clock researcher, asked how light synchronized the body clock. But before I elaborate on why Foster's findings were so important (far beyond clock research), I will have to take another detour.

Children and clock researchers have a common interest: hamsters and mice. The daily activity rhythm of most rodents can be easily recorded because they love nothing more than running in a wheel. Not only "bored" laboratory rodents, living in captivity, become wheel-running junkies. The story goes that an American clock researcher stored unused running wheels from the lab in his garage.

Coming home late one evening, he heard the familiar squeaky sound of turning wheels coming from the garage and found that a family of wild mice had adopted the garage as a workout gym. Most rodents are active at night and sleep during the day, so they run at predictable times every day or rather every night.[4] To record a rodent's body clock, we simply have to hook up the shaft of the running wheel to a little electric contact that is closed once every rotation. In the early days of clock research, the switches were connected to a pen mounted on a slowly turning roll of long paper. Nowadays, the impulses from the electrical switches are recorded by computer programs. The original recording set-up produced broad bands of ink when the pen was jerking up and down each time the contact closed—that is when the rodent was running in the wheel—and a straight, thin line when the animal was asleep. If the paper was moving at a constant speed (for example, one centimeter per hour), all we had to do was to cut the paper strips into pieces of twenty-four centimeters and paste them underneath each other onto a piece of cardboard to produce data like that shown in this chapter.

The first graph shows the results of an experiment that lasted thirteen days.[5] Each strip, one above the next, represents the animal's wheel-running activity during one day. Since each centimeter on each strip is equivalent to one hour of the day, one can easily see at what times the animal was running in the wheel. The animal started to run at approximately the same time every day, so that the onsets of the broad ink bands fall onto a straight vertical line. In this case, the body clock produces days of exactly twenty-four hours, and you can probably guess by now that in this example the animal was kept in a light–dark cycle. If the activity of the bunker subject, who lived through twenty-five-hour days, were recorded in a similar way (there were contacts in the bunker's floor), the data would look quite different.

On average, the subject's activity starts an hour later every day, falling on an invisible oblique line tilted to the right. The disadvantage of plotting the data this way is that the regular bouts of activity are split into two components once they run into the edge of the cardboard. This is why clock researchers use "double plots." In the early days, researchers would photograph the original cardboard and develop two prints that were glued next to each other on a new piece of cardboard, the right half being displaced upward by one day. In this way they would be able to see all activity bouts as continuous broad bands.

If we would plot the activity of a bunker subject whose internal days were shorter than twenty-four hours, the activity onsets would fall on a line tilted to the left since he or she would get up earlier every day (in reference to local time). This is actually what mice do when they are kept in a dark, time-free environment. The final graph

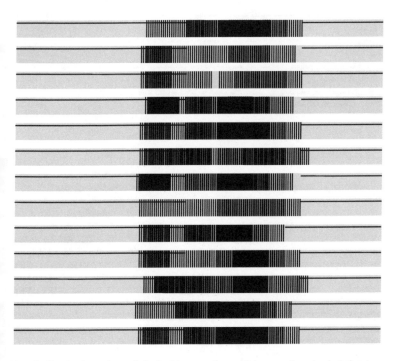

A rodent's wheel-running activity. In this case, the rodent ran on the wheel at about the same time every twenty-four-hour day.

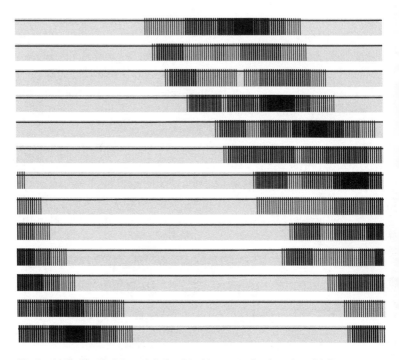

The bunker subject's daily activity level for his twenty-five-hour internal days.

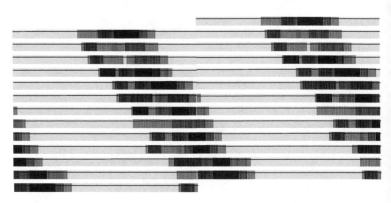

A double plot shows bouts of activity as continuous bands without interruption by midnight.

shows an experiment recording the activity of a mouse that was kept under a light–dark cycle during the first seven days (indicated by the black and white bar at the top of the graph) and was then released to constant darkness, whereupon its activity rhythm began to run free.

You can see that the activity on the first day in constant darkness makes a little "jump to the left." This is not because the subject's body clock suddenly jumps in time but because a mouse simply doesn't like to run in the bright laboratory light, even if its body clock "tells" it to be active. As soon as the lights go off, however, the mouse starts to run.[6] To estimate where the activity would have begun if it hadn't been suppressed by light, clock researchers look at the rhythm when the clock was released to constant conditions and follow its path back to when the animal was still in a light–dark cycle (see the dashed lines). In this mouse's case, this method shows that the clock's

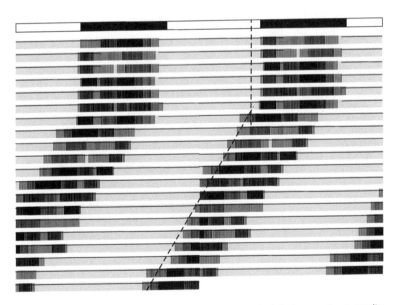

This rodent was released from a light–dark cycle to constant darkness, whereupon its activity rhythm began to run free (at the intersection of the dashed lines).

signal to start running came approximately one hour before lights went off.

At last, you are equipped with enough information to readily understand the results and conclusions of Russell Foster, whose experiments approached the problem of how light synchronizes the body clock by a very straightforward question: which receptor type translates the information about light and darkness to the clock? He used mouse strains that have no rods and submitted them to experimental conditions much like those shown in the last figure. The results were easy to interpret. These mice could be synchronized just as well by light–dark cycles as mice with a perfectly intact retina. So rods were apparently not the receptors informing the clock about night and day. He then took mouse strains that had no cones and again found that these animals synchronized as well as intact or rodless mice. Once more, the results were easy to interpret. The clock apparently used both rods and cones to "know" about light and darkness. Without rods, it used cones; without cones, it used rods. So all that Foster had to show was that the clock of mice lacking both rods and cones could not synchronize to light–dark cycles. But when he tested mice lacking both receptor types, he was in for a big surprise: the activity recordings were essentially the same as in the previous experiments.

But maybe the clock doesn't even need the eyes to "see" the light. Perhaps it uses some other light-sensitive tissue—for example, the skin.[7] Around the same time Foster performed his experiments, another clock researcher showed that one could set the human clock by shining light onto the back of the knee. But for several years, other clock researchers failed to replicate these results, clearly demonstrating that the clock only got light information through the eyes. This was already well known from numerous animal experiments—the clock in eyeless animals didn't synchronize to light–dark cycles. Thus, the only conclusion from Foster's experiments was that there must be another, hitherto unknown light receptor in the eye.

This hypothesis raised a big red flag in the community of eye researchers, who thought that everything one could possibly know about the eye of mammals was already known. Many ophthalmologists refused to believe that an unknown light receptor had been discovered by a clock researcher. A couple of years later the enigmatic receptor was eventually identified. When the Foster experiments were eventually repeated with mice lacking all receptor types—rods, the three cones, and a new receptor—their clocks finally stopped synchronizing to light–dark cycles.

The newly discovered light receptor is related to a receptor that allows amphibians to rapidly change their skin color. Rods and cones are actual cells embedded into the retina. The conversion from light to cellular signals is a biochemical process involving proteins that can catch photons (light "particles") with the help of pigments, such as carotenoids in red carrots or chlorophyll in green leaves. Carrots look reddish because their carotenoids absorb blue light, so that the remaining, reflected light appears red, whereas leaves absorb both blue and red light and therefore look green. Both rods and cones use the same type of protein, called *opsin,* but the opsins in rods and in each of the cone types are slightly different, absorbing light of different wavelengths, and thus enable us to identify colors. The new photoreceptor serving the mammalian body clock is also an opsin. Because it is part of those amphibian cells (melanophores) that enable the skin to turn dark rapidly, the new opsin receptor was called *melanopsin.*[8] Light reception by melanopsin does not use specialized cells such as rods and cones, which collect information about the exact where and when of a light stimulus. Melanopsin is spread widely across the retina in those nerve cells that give rise to the long extensions that bundle together to form the optic nerve.

When scientists want to fathom a biological phenomenon, they of course want to know where the phenomenon (or its control) is localized. The question about the localization of the body clock was pursued extensively in the 1970s, and the outcome depended very much on the species investigated. In plants, a distinct localization was never found. The clock appears to be everywhere—in leaves, stems, and roots. In animals, the clock was found in the brain. In insects, it resides in a small number of highly specialized neurons; in slugs, in neurons at the base of the eye. In reptiles and many birds the clock resides in an annex of the brain called the *pineal*.[9] The pineal is a gland that produces the hormone melatonin, which serves as a signal for darkness.[10] In birds, amphibians, and reptiles, the pineal gland is sensitive to light, which it receives directly through the skull. In mammals, the pineal has lost its ability to perceive light but still produces the hormone melatonin at night (notably, in both night- and day-active animals). Melatonin makes us (and probably all other day-active animals) sleepy and a bit colder, while the same hormone has the opposite effects in nocturnal animals. But melatonin does not act just like a normal sleeping pill; it can also affect the body clock so that a regular prescription can synchronize its rhythm. You may have wondered how we can help people like Harriet in her existence as a periodic shift worker. Melatonin is certainly an option and has been used successfully in synchronizing the body clock of totally blind people to the twenty-four-hour day.

In mammals, the clock was found in a small group of neurons above the optic chiasm. A small brain area dedicated to a specific function is called a *nucleus,* and since this clock-nucleus is located directly above the optic chiasm, it is called the *suprachiasmatic nucleus* (SCN).[11] The SCN is quite a remarkable little piece of tissue. With no more than approximately twenty thousand cells, it appears to contain all that is needed to make the body clock tick—to keep internal time.[12] If the SCN is removed from a rat or a hamster, its wheel-running activity immediately loses its daily regularity. If an SCN is implanted into the brain of an animal that had its own SCN removed, its wheel-running activity becomes rhythmic again with the expected circa-twenty-four-hour period.[13] The SCN is the main center of the body clock not only in hamsters, rats, and mice but also in humans. Patients with lesions in this brain area have great difficulty in keeping to regular sleep–wake and activity–rest schedules.

I have covered a lot of ground in this chapter, beginning with the sleeping difficulties of a blind woman, then discussing the discovery of a new light receptor, specialized not for visual tasks but for the nonvisual task of generally sensing night and day, and finishing with the location of the clock's centers. But there is still an open question—the one that Harriet asked herself. Why does the body clock of some blind people remain synchronized to light and dark while that of other blind subjects runs free? We still don't know the full answer to this question, but we do know that the body clock of people who have no eyes almost always runs free. If eyes lack rods and cones but can still unconsciously sense light by an intact melanopsin system, the clock can still be synchronized. An unconscious light perception may seem strange, but it does exist. Russell Foster investigated blind subjects who had no conscious light experience but whose body clock was still perfectly well synchronized. To identify this unconscious light perception, he presented them with light stimuli accompanied by a short sound. Sometimes (randomly), he also presented the sound alone, without an accompanying light stimulus. He encouraged his subjects to simply guess, even if they had no conscious experience of light, whether or not they sensed the light when they heard the sound. The results of these experiments showed that the subjects "guessed" significantly better than random. Their brain must have registered the light even though they did not consciously experience a sensation.

The clock of individuals whose eyes lack rods, cones, and melanopsin tends to run free, but some apparently remain synchronized. The reason for this may lie in the fact that the internal days of these blind people are already very close to twenty-four hours, so that other signals—activity or meals that are normally too weak to set the clock—might be strong enough to "nudge" it into synch with the outside world. Although Harriet in this chapter's case still has eyes, she cannot perceive light via any of the receptor types. Her free-running body clock is about one hour slower than the rotation of earth, so her internal time is not synchronized to her social schedules, and she has to live her life as a perpetual periodic shift worker.

7

A new shipment of hamsters from the Charles River Breeding Laboratories had arrived. Christopher was in the process of cataloguing the new animals. When he had entered them all into his files, he placed each one of them into a separate cage equipped with a running wheel. His first task with newcomers was to record their circadian activity rhythms—a routine to check whether they were good or bad runners. A couple of weeks later, he sorted through the pile of activity recordings and made his usual list: "good runner" or "bad runner." A typical activity recording for hamsters living in constant darkness, one which Christopher had seen hundreds of times before, shows a free-running rhythm with a period close to twenty-four hours.

Halfway through the pile of double plots, he caught his breath. Hamster #31M18, a male, showed a very unusual activity pattern— different from every other hamster activity recording he had seen. The period of this hamster's clock was only twenty-two hours! Christopher immediately went to the constant condition chamber and opened the box of four hamster cages that contained the label #31M18. He wondered whether the animal was sick and was even prepared to find a dead hamster in the cage. Hamster #31M18 appeared to be sleeping, so he took it out of its cage and gently stroked its fur. It was a pretty sleepy hamster, but it did not appear to be sick. After placing it back in its cage and closing the lid of the box, he went to check the timer of channel 361, which was responsible for recording the wheel-running activity of the unusual specimen. He ran some tests, but the timing of the recording device appeared to be normal.

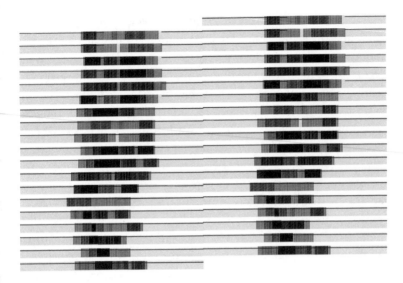

A typical activity recording for hamsters living in constant darkness—a free-running rhythm with a period close to twenty-four hours.

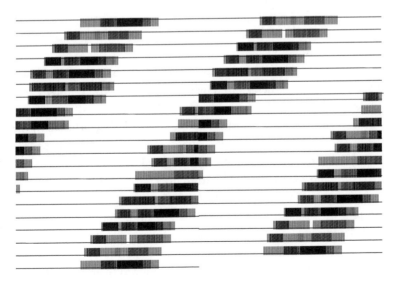

Hamster #31M18's double plot shows a period of only twenty-two hours.

Christopher went back to his office and configured his computer to display the activity of channel 361 online. The double plots he had looked at earlier were already a couple of days old, so he checked the recording of the last couple of days and found that the unusually fast activity rhythm of hamster #31M18 had persisted up to now. The animal's tiredness was to be expected because it had stopped using the wheel only a couple of hours earlier, which meant that it must have been in deep sleep when he had taken it out of its cage. For the rest of the day, Christopher couldn't concentrate on his routine lab work, pondering the short circadian period of the newcomer. In the late afternoon, at the daily group meeting with the head of the lab, he waited until all of his colleagues had presented their reports before pulling out the double plot of hamster #31M18. As he had anticipated, all participants were clearly astonished, and he thoroughly enjoyed being the center of attention. The afternoon meeting took much longer than usual and produced several important decisions. The behavior of the unusual newcomer was to be tested under a light–dark cycle (the group decided to simulate a summer day of fourteen hours of light and ten hours of darkness). After three weeks, the hamster would be transferred to the breeding facility to mate with three females that showed the usual hamster rhythms, very close to twenty-four hours in constant darkness.

The outcome of the recordings in a light–dark cycle showed that hamster #31M18 synchronized perfectly. There was, however, one striking difference: the newcomer started his days about four hours earlier than "normal" hamsters. Several weeks later (fortunately, the pregnancy period of a hamster is only about seventeen days), the first offspring could be tested both for the speed of its clock in constant darkness and for its ability to entrain to a light–dark cycle. It turned out that half of the offspring had inherited the clock of their father and the other half that of their mother. Of those that resembled their father, only a few were able to synchronize properly. For several days, they started their activity in synchrony with lights turning off (hamsters are nocturnal, active at night) but then suddenly got up earlier every day by several hours, as if their body clock were ignoring the light–dark cycle, thereby running free with an abnormally fast rhythm.[1] Once their activity again came close to the dark portion of each cycle, it seemed as if they were able to synchronize to the light–dark cycle for a couple of days, but then they broke away again, displaying their very short rhythms. Few of the offspring that had such short activity rhythms ever synchronized and—like their father—they began their days about four hours earlier than "normal" hamsters. The team proceeded to do more breeding with the offspring. When they eventually crossed two animals with a fast clock, they discovered that all of the resulting offspring had an even faster clock than their unusual ancestors.

As much as biologists like to know the anatomical location of the phenomenon they investigate, they also want to know which genes are involved in its function. Scientists have only recently dared to look for genes that are responsible for biological qualities (traits) that have more complex genetics than skin or hair color. The biological clock is certainly considered such a complex trait, and for exactly that reason, one of the pioneers of clock research, Colin Pittendrigh, was initially quite skeptical about whether scientists could identify genes that were involved in the biological clockwork. Seymour Benzer was one of the first scientists who dared to look for genes responsible for complex traits, and he met with a lot of skepticism when he approached these questions relatively late in his already successful career.[2] Together with his student Ron Konopka, he set out to look for "clock genes." They found that mutations in a single gene—which they called the *period* gene—drastically changed the free-running rhythm of the fruit fly's clock.[3] Mutations at different locations within the same gene produced flies that lived either very short or very long days in constant conditions. Other mutations even rendered the flies arrhythmic. Shortly after the discovery of the first clock gene in the fruit fly, genes with similar properties were identified in other organisms. The next in line was a bread mold.[4] The identification of clock genes in mammals took another two decades, however.

The discovery of clock genes in the fruit fly and the bread mold had been known for almost twenty years when Christopher discovered the unusual hamster.[5] The excitement of the team about their new discovery was, therefore, not surprising. Because hamster clocks are usually very accurate and reproducible with a period close to twenty-four hours in constant darkness (hamsters very rarely live days shorter than 23.5 hours), the abnormal rhythm of hamster #31M18, which they called the *tau* hamster, suggested a mutation in a clock gene.[6] When genes are involved in a biological function, we can apply classical genetics to help us understand more about the underlying genes. This always involves breeding, which is why the scientists in our story decided to do breed the *tau* hamster and its offspring.

The outcome of these experiments proved that the fast, free-running rhythm had a genetic basis. The combination of offspring suggested that hamster #31M18 carried only one gene with a *semi-dominant* mutation while the other gene appeared to be "normal."[7] If the newcomer carried one normal version of the gene (N) and one mutated version (M), and if his mate carried two normal copies of the gene (NN), the combination in their offspring was predictable. Since each of them can only inherit either M or N from their father and only N from their mother, the chances of inheriting an MN or NN are equal. Half of the offspring in the first generation should show the same period in their wheel-running activity as their "normal" mother and the other half should inherit the fast clock from their father. This was in fact the outcome of the first round of breeding experiments with hamster #31M18.

In the second generation, however, the dice are thrown again. This time it is possible to mate two MN animals—hamster #31M18's offspring. So after several rounds of breeding, the team took two of the offspring with the fast, twenty-two-hour free-running rhythms of their father. Again the outcome is predictable. M and N now have equal chances, so that 25 percent of the next generation should have NN (the normal clock of their grandmother), and 50 percent should have MN or NM, showing the fast clock of their grandfather. The remaining 25 percent, however, should carry MM, and if one mutated gene could shorten the period of the rhythm by two hours, animals carrying two of these genes may have an even shorter period, which turned out to be true. The *homozygous* animals live days as short as twenty hours.[8]

Twelve years later, proof was provided that Martin Ralph, alias Christopher, had discovered the first known mammalian clock mutant. The fast clock of hamster #31M18 was due to a mutation in a single gene that was crucial for the clock to tick at its normal rate. Experiments with this mutant also underlined the important role of the suprachiasmatic nucleus or SCN in the mammalian body clock. The transplantation experiments described in the last chapter showed that the SCN was able to rescue the rhythmicity in arrhythmic animals. Martin Ralph performed similar transplantation experiments with the *tau*-mutant and the normal hamster. The recipients of an SCN removed from a *tau*-mutant hamster showed a fast activity rhythm in spite of the fact that their genome should have produced a normal activity rhythm. The opposite was also true: mutant hamsters lived normal circa twenty-four-hour days if they received an implant from a normal hamster (supporting the role of the SCN as the master clock in mammals).

Thus, the clock mechanism involves dedicated genes that determine the clock's qualities, such as its free-running period. But the *tau*-mutations apparently also changed the relationship between internal (body clock) time and external (light–dark cycle) time. Under constant conditions, the shorter their period, the earlier the mutant hamsters rose to start their day.

8

It was 5 A.M. when the family entered their 24/7 work-out facility in Utah. A gray day with a slight drizzle had just begun. The large, cheerful group was quite a sight. There was great-grandmother Sarah; her daughter and son, Alicia and Frederic; Alicia's son and daughter, Phillip and Rebecca; Rebecca's two teenage girls, Julia and Anna; and Frederic's two youngsters, Peter and Jessica. It was almost the entire tribe. Only Sarah's two sisters, Isabelle and Judith, were missing. They had arisen with the workout group but had stayed at home, responsible for the traditional family breakfast later on, which would last for hours. Alicia's, Frederic's, Rebecca's, and Judith's spouses and Judith's children were still sleeping soundly —the "early shift" had left the house on tiptoe to avoid disturbing the "late shift" of the family.

Sarah was in seventh heaven. During the summer holidays the extended family always came together, and she thrived in the presence of her children, grandchildren, and great-grandchildren. She had had her first child, Alicia, at the age of nineteen and was only twenty-one when Frederic arrived. Early marriages ran in her family. When Alicia made her a grandmother she was only thirty-nine, and her first great-granddaughter was born shortly before her fifty-eighth birthday. Now, at seventy-five, she didn't exercise with the rest of her family but refused to stay home when they all went on their traditional dawn gym outing. She sat on the long bench on one of the sides of the large workout room watching the various activities of her offspring. Alicia and Frederic, her children, were chattily jogging next to each other, occupying two of a long row of treadmills. Alicia was still in great shape at fifty-six and liked to gently mock her younger brother, who puffed quite a bit on his running machine. He could do with some exercise because he was already showing quite a belly. Granddaughter Rebecca was using the rowing gear without exhausting herself too much while watching her teenage girls, who spent more time giggling than hitting the punching balls at the other end of the gym. Rebecca had always taken life easy, unlike her brother Phillip, who was incessantly hard on himself. "If he hadn't such ridiculously high standards, he might be married by now," Sarah thought. Being single at the age of thirty-three came close to skipping a generation in her family—his much younger cousins already had had steady partners for some time now, and Sarah hopefully anticipated the arrival of more great-grandchildren.

By seven o'clock, they were all back in the large kitchen sitting down to the breakfast the great-great aunts had meanwhile prepared. Julia and Anna were still lively. They were probably making fun of their uncle Phillip again—it seemed to be their favorite pastime. "Quiet down a bit you two, daddy and the others are still sleeping," Rebecca admonished them without much of an effect. Alicia's husband, Terry, was the first of the "late shift" to arrive at around eight when they were already sitting in front of their empty breakfast plates. He was served coffee by his mother-in-law. Within the next hour, all members of the big family were sitting around the extensive kitchen table, happily planning the rest of the day.

The sequence of appearances or rather disappearances was reversed in the evening. After their usual six o'clock dinner, the two ends of the generation scale played games in the drawing room while the rest were reading or engaging in discussions on the porch. Sarah and her sisters were the first to retire at around 9 P.M., about an hour later than they normally did when the big house wasn't filled with family. No matter how late they went to bed, they would still wake up around 4 A.M. without being able to go back to sleep. Her family had always been on the early side, as she remembered, right back to her grandparents. Before Sarah went to bed, she took the medication for her blood pressure problems that she had forgotten about earlier when she was concentrating on her card game. "Take it an hour before bedtime" were the doctor's orders. She wondered whether he meant her usual bedtime or the actual time she went to bed.

Not much later, the other members of the "early shift" said good night one by one. Great-great aunt Judith and her son (his brother had died some years back) as well as her two grandchildren were the last of the "early shift" to hit the hay. The remaining spouses were left behind, smiling to each other about the strange family they had married into and enjoying their evening for another couple of hours in peace and quiet. Judith's husband, James, and Frederic's wife, Mildred, were the last to go up. They had spent the evening exchanging anecdotes about the early shift. James insisted that his part of the family was not quite as bad as the others. On her way up, Mildred made herself some chamomile tea. While she was waiting for the kettle to boil, her tired gaze was caught by a drawing, which Sarah had recently framed and hung on the kitchen wall. She looked at the various circles and squares, smiled, and thought that James was probably right.

You may have recognized that the "early shift" of the Utah family has some relationship with hamster #31M18 and its offspring in the previous chapter. The mutant hamsters with their fast clock started to run in their wheels in the afternoon—four hours earlier than their "normal" litter mates. The visit to the gym is somehow comparable to wheel running; the "early shift" in Sarah's family starts its day before sunrise, much earlier than other humans. Thus, to be earlier or later than the other members of one's species is independent of being night-active, like hamsters, or day-active, like humans. The "early shift" in Sarah's family suffers from what is technically called *Advanced Sleep Phase Syndrome* and—as in the case of the *tau*-mutant hamster—the affected individuals of the Utah family have a gene that carries a mutation. It is not the same gene that is responsible for the short days of hamster #31M18 and his offspring, but the functions of these two genes are closely linked. Metaphorically, the hamster mutation produces a blunt screwdriver that cannot turn the screw efficiently whereas the mutation in Sarah's family makes a blunt screwhead so that the screw cannot be turned efficiently, even by an intact screwdriver. In the *tau*-hamster, the affected gene encodes an enzyme that modifies other proteins, among them an important clock protein.[1] With its mutation, the enzyme doesn't do its job as efficiently as it normally does. The gene that carries a mutation in Sarah's family, on the other hand, encodes the important clock protein so that it cannot be modified as efficiently.[2] Thus, although the mutations affect different genes, they end up having the same effect.

I have gone into the molecular and genetic mechanisms of the body clock in some detail, although they may not be essential for understanding the phenomenon of internal time. I've done so because, in my experience, some people still think the body clock is something esoteric rather than a profoundly biological function. When I joined the medical faculty in Munich many years ago, I followed the custom of introducing myself to my colleagues. Several of them were directors of large clinical departments. Although all were extremely friendly, some of them didn't hold back their opinion about the circadian clock. One of them even said, "My dear colleague, this is all fascinating, but surely the biological clock is only an issue for very sensitive people." Yet the biological details, right down to the molecular and genetic levels, prove how much biology is behind our internal timing system and how well our field already understands its functioning.

The last chapter demonstrated that one can find dedicated genes which are essential for the body clock to tick properly and that their function within the clock appears to be the same (conserved) across very different species—from insects to humans. Some of the molecular geneticists who were instrumental in discovering the first clock genes were naively optimistic, hoping they had "cracked the clock" when the first understanding evolved of how a very limited set of genes and their protein products come together to generate an internal day—and so were the media.[3] The concept of the molecular clock was ingenious and simple. At the Gordon Research Conference for Chronobiology in Irsee, Germany, in 1991, Michael Rosbash, who presented the results of his student, Paul Hardin, based on experiments in the fruit fly, opened his talk with the following joke, which I relate from memory.

A world-famous physicist traveled the country to give public speeches about his groundbreaking work. One evening, on the way from a city where he had just finished his presentation to the next town where he would be giving the same speech all over again, he remarked to his chauffeur how sick he was of telling his story for the umpteenth time. The chauffeur looked at him in the rearview mirror and smiled gently. "I can understand your frustration, Sir, because meanwhile even I know your speech by heart." "Does that mean that you could actually GIVE my speech?" replied the professor, and the chauffeur simply said, "I believe so, Sir." "Then why don't we switch roles tomorrow evening to give both of us a break—it's a small town and nobody knows my face." So the next evening, the chauffeur, dressed in the professor's suit, climbed onto the stage while the professor, wearing the chauffeur's uniform, took a seat in the last row. The fake professor did a superb job. He really knew every word of the speech by heart and even produced the little jokes with perfect timing. He also mastered all the questions in the discussion that followed as if he had done the famous experiments himself. At the end of the discussion, however, a physics professor from the local college asked a difficult question, one that had never been asked before, so the chauffeur couldn't know the answer. He paused for a fraction of a second and then replied with a broad smile. "The answer to this question is so simple that it can even be given by my chauffeur," he said, pointing with a grin to the person in a chauffeur's uniform sitting in the last row.

At the climax of his introductory joke, Michael Rosbash pointed at Paul Hardin, who was sitting in the back of the room, as if Michael was the chauffeur and Paul the professor. He then proceeded to explain their hypothesis of how molecules generate a circa twenty-four-hour rhythm. The hypothesis describes a simple negative feedback loop. Think of a chain-production process that produces the most delicious pralines as its final product. Worker number 1 is a very self-centered and vain person who is extremely proud to be one of the most important people in the production process. He is responsible for getting the secret recipe out of a safe. He then copies it onto a piece of paper, which he hands through a small opening in the wall of his room to another worker. Worker number 2 then produces the heart of the praline and diligently throws away the piece of paper so that the recipe cannot be stolen. The unfinished praline is now handed to worker number 3, who is responsible for its icing and other delicious decorations. Once a considerable number of the delicious pralines have been made, they are handed back to worker number 1 for inspection. However, as soon as our self-centered, vain recipe-copier sets eyes on the first final product, he stops everything and merely looks with pride at the praline. No more copies of the recipe are handed through the little hole in the wall. No more pralines are produced and refined with icing and decoration. Finally, a fourth worker collects the finished pralines to be shipped off to customers.

In this scenario, the production line would stop if worker number 4 were not to collect the finished products. But as soon as he has removed the last praline, worker number 1 wakes up from his spell and continues to copy the recipe on little pieces of paper, so that the production is reinitiated. As a consequence, the amount of pralines being produced oscillates in an endless rhythm.

So much for the metaphor—now to the scientific explanation of how cells produce circadian rhythms with the help of genes and proteins.[4] In the nucleus, the DNA sequence of a clock gene is transcribed to mRNA; the resulting message is exported from the nucleus, translated into a clock protein, and is then modified.[5] This clock protein is itself part of the molecular machinery that controls the transcription of its "own" gene. When enough clock proteins have been made, they are imported back into the nucleus, where they start to inhibit the transcription of their own mRNA. Once this inhibition is strong enough, no more mRNA molecules are transcribed, and the existing ones are gradually destroyed.[6] As a consequence, no more proteins can be produced and the existing ones will also gradually be destroyed. When they are all gone, the transcriptional machinery is not suppressed anymore, and a new cycle can begin.

This negative feedback loop hypothesis was created at a time when only one clock gene was known in the fruit fly. A single gene and its protein were presumed to create the internal day simply by acting on each other in a negative feedback loop. The authors, however, predicted in their first paper that more players were bound to be found. Shortly after the birth of the negative feedback hypothesis, another important clock gene was identified in the fruit fly, and now we know more than twenty genes to be involved in building the body clock of animals: the picture of the molecular circadian machinery has become rather complex. The idea of negative feedback loops is still the basis of the current hypotheses that try to explain how cells generate daily rhythms, but many components are thought to form a network of feedback loops. Despite this complexity, the important take-home message is that daily rhythms are generated by a molecular mechanism that could potentially work in a single cell, for example a single neuron of the SCN.

The fact that many genes are involved in the biological clock predicts that many things can go wrong by mutation. There can, therefore, be many causes for somebody's body clock ticking differently from other individuals, making him or her much earlier or much later than the rest of the population—far beyond the two genes we have encountered in the *tau*-hamster and in Sarah's family. The picture Mildred looked at while waiting for the water to boil is similar to a family tree made by the scientists who investigated the phenomenon of the "early shift" and its genetic basis.[7]

In this family tree, you may notice the dotted square surrounding the descendants of Judith and James. Great-great aunt Judith and her descendants are also early birds, but not quite as early as Isabelle or Sarah, or Sarah's children and grandchildren. Although the geneticists also identified some members of Judith's branch of the family as early birds, these family members did not carry the mutations found in the others. The family tree that Sarah was given by the geneticists shows only part of the even bigger family tree, of which many branches show that not every individual in the family inherits the life-pattern of a lark from his or her respective parent—genetics is hardly ever a simple yes-or-no story.

Being a lark or being an owl definitely has a genetic basis, but the

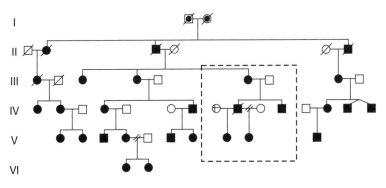

Sarah's family tree, genetically expressed. Circles represent women, squares represent men; filled circles and squares represent affected individuals, empty circles and squares represent unaffected individuals. Generations are represented by the roman numerals on the left. Redrawn from K. L. Toh et al. (2001). An h*Per2* phosphorylation site mutation in familial advanced sleep phase syndrome. *Science* 291(5506):1040–3, with permission from *Science*.

genetics are complex—which is no surprise considering the number of genes that are involved in building body clocks at the molecular level. You have read here about clock genes and their products, the clock proteins. Beyond these clock genes, there are many other genes that make people early or late; for example, those that convey light information to the clock, such as melanopsin.

The circadian system is not merely a neuronal center that ticks away in our brain. It is an entire pathway with inputs, clocks, and outputs. This pathway starts by detecting light or darkness in the eyes and sending this information to the clock neurons in the SCN. These in turn inform other parts of the brain (for example, sleep centers) or organs like the liver about what *internal* time of day it is. All the components of this pathway are important in determining our chronotype.

The Elusive Transcript

9

To his great frustration, Oliver's experiments had utterly failed for several weeks now. He was working on a project that concerned the biochemical and molecular regulation of liver metabolism, continuing the experiments of a Canadian postdoctoral fellow who had recently left the lab for his first job as a professor and scientist.[1] The experiments focused on a gene that was switched on in liver cells under certain conditions. Although Oliver had always meticulously followed the experimental protocols, he was never able to find the products that should have been present if the gene had been activated. Oliver had always been an excellent student with considerable lab experience. He had done a superb job in the experimental work that eventually had led to his final thesis. For weeks he had been trouble-shooting his procedures, but he couldn't find the reason why his predecessor had always found huge amounts of the expected gene products while he drew a blank every time.

It was approaching noon. Oliver had been in the lab since six in the morning, constantly working at the bench. As the son of farmers, he was used to getting up very early and loved being in the big lab all on his own before the others arrived. That way he could fully concentrate on his experiments while listening to his favorite music (some of his co-workers who populated the lab later in the day didn't share his taste). Now the lab was full of his team members, but in spite of his exhaustion, he decided to do another extraction of liver tissue. After several hours of grinding, extracting, measuring optical densities, pipetting, amplifying, loading gels, and running them in the electrophoresis set-up,[2] he was back where he had started. The expected gene products were undetectable.

It was getting late. He was almost alone again in the lab, looking through the animal records and checking once more whether there was a difference in the rats that had been used by his predecessor compared with those he sacrificed to harvest liver tissue. But as he expected, he did not find any indication that this could be the source of his problems. He began to hate the ghost of the guy who had done the initial successful experiments and who was quite a legend in the lab, one of the few who had received a *summa cum laude* for his thesis. The experiments Oliver was trying to replicate were only a side project that was left unfinished when the Canadian had moved. His daily habits had apparently been the opposite of Oliver's. He hardly ever arrived in the lab before noon, and he experimented throughout the night—until almost the time Oliver usually arrived for his day's work.

Oliver went out to get some fast food with a lot of coffee and decided to try his luck one last time. He would start from scratch, sacrificing a new rat and going through all the experimental procedures again before he confronted his professor with his long line of negative results. He was certainly not looking forward to this meeting, especially since his negative results contradicted those of a star experimenter. He was just a Ph.D. student, and the other fellow had been a longtime postdoc. Either he was a complete loser, or his highly praised predecessor had made a big mistake—both possibilities would be difficult to communicate to his boss. But, on the other hand, he was sure that he had made no mistakes in his experiments, so the other guy must have blundered. He was still frantically working through the laborious protocols when his coworkers started to arrive. They were not astonished to see him—they were used to Oliver already being there when they came to the lab. But since they had left him the evening before, still working hard, they all knew that he must have pulled an all-nighter. When Oliver had finished loading his gels and had started the electrophoresis apparatus, which would take several hours to run, he finally went home to sleep, making sure his blinds were shut as tight as possible so as not to let in any daylight.

He woke up at noon, showered, got dressed, and made himself the first real meal he'd had for almost thirty hours. He desperately wanted to get to the lab. He almost got run over by a car on his way because he could only think of the outcome of the experiment he had started earlier that day. When he finally arrived at the institute, he threw his knapsack onto the desk in his small cubbyhole and went to check his gel. To his great surprise and relief, it showed a large amount of the expected gene product. He went out for a long walk, trying to make sense of why he could sometimes see the wretched protein and sometimes not. What had he done differently? How could all this make sense? Suddenly he stopped, turned around, and jogged back to the institute. He had come up with a hypothesis and had thought of how to verify or falsify it—the best kind of experiment a scientist can think of. He was glad that he had slept because the experiment he was now planning would take more than twenty-four hours—and then at least another day would be needed to do all the preps and analyses. Two-and-a-half days later he sat at his desk completing his lab book. Finally, he pasted the photograph of his last western blot on the last page of his notes, looked at it for a long time,

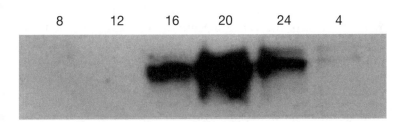

Oliver's western blot, showing the protein of interest. This western blot is the work of the real "Oliver." Reprinted from J. Wuarin and U. Schibler (1990). Expression of the liver-enriched transcriptional activator protein DBP follows a stringent circadian rhythm. *Cell* 63: 1257–1266, with permission from Elsevier.

and smiled.[3] Suddenly, he was very much looking forward to the meeting with his professor.

⧗

The discovery of the SCN as the clock-center in mammals was a huge step toward understanding internal timing. This small group of SCN neurons appears to contain all the necessary features required to control the daily events of our body. It receives light information via the optic nerves from the eyes and can thereby synchronize to day and night. If it is lesioned, the regularity of the daily activity rhythm is disrupted. What else do we need in order to create an internal timing system?

So far, you have read about the daily changes in our behavior, about when we sleep or are active, about the fact that our body's temperature swings from a low in the late night to a high in the evening.[4] Together with the curious astronomer de Mairan, you discovered the daily movements of plant leaves. Surely, the clock has not evolved only to control these few qualities in the life of an organism. We know from experience that there are many more functions that change over the course of the day: at certain times, we prefer to exercise; at others, we are hungry; and our preferences for food change drastically from morning to evening.[5] The times when we can best concentrate, solve puzzles, and speak foreign languages also change over the course of the day. At lunchtime, we react to alcohol differently than in the evening. Statistics even show that people prefer certain times of day to have sex. Other studies reveal that the pain we suffer when going to the dentist is not the same throughout the day. Road accidents are highest at around 4 A.M.[6] Many of us experience a postlunch dip and would like to have a little power nap (or if we lived in Mediterranean regions, we would have an extended siesta). Many kinds of medication should be taken at a specific time of day (recall Sarah's thoughts in the last chapter about her blood pressure pills). You probably have encountered the medical routine of asking patients to fast before coming to the clinic at eight in the morning for a blood draw. Many components in the blood go up and down over the course of a day. The hormone cortisol is a good example of such a component. Its concentrations in the blood are highest in the morning, decline with some wiggles over the course of the day, reach their minimum during the first half of the night, and then start to rise again toward their morning peak. Although cortisol concentrations depend on many factors, their daily rhythm is controlled by the body clock. So if we measure its concentration at 8 A.M., the results will very much depend on the patient's chronotype. The science that investigates when to take what medication (chronopharmacology) is a very important branch of clock research. Its aim is to find the right time of day for a drug—when a minimal concentration exerts the maximal effects—thereby also reducing any undesired side effects of the drug. Again, the optimal time of a drug will vary greatly with the patient's chronotype.

The aim of this chapter is to show how profound the influence of circadian timing is. The simple reason for Oliver's initial failure to replicate his predecessor's results had to do with his daily routine.[7] While he, the doctoral student, arrived in the lab well before the rest of the team, sacrificing rats for the extraction of liver tissue in the early morning, the Canadian postdoc came to the lab late and worked well into the night, sacrificing the rats for his experiments in the evening. The gene they investigated was only activated toward the end of the rat's sleep phase, which, in a nocturnal animal, corresponds to the late afternoon of our day.[8] Oliver came to this conclusion on his long walk and planned an extensive experiment in which he would sacrifice rats every four hours over the course of twenty-four hours and prepare the respective extracts for the detection of the binding protein, DPB. The results clearly verified his hypothesis. The first time point at which he could detect the protein was at 4 P.M. At eight the protein was even more abundant, but it was already declining around midnight, becoming undetectable again at 4 o'clock in the morning. No wonder Oliver was unable to detect the protein whereas the Canadian postdoc found bucketloads whenever he made an experiment.

The consequences of this discovery were substantial but retrospectively not really surprising. This is true for so many important findings—once we have been made aware of them, they make so much sense in the context of what we already know. Oliver's findings indicated that the circadian clock apparently controls the activation and deactivation of individual genes. Many of the metabolites collected in the Andechs bunker experiments went up and down in synchrony with the internal days that subjects experienced in their time-free existence.[9] Clearly, metabolism is under the control of the circadian clock. The regulation of practically all functions involves, directly or indirectly, the regulation of genes—they are the templates for the cell's tools that organize metabolism.[10] So it is not surprising that the circadian clock also "uses" the regulation of gene expression to organize metabolism appropriately within the twenty-four-hour day. But still, this had to be shown, and Oliver's results showed how profoundly cells use gene regulation as a circadian mode of regulation. Depending on the tissue, 15–40 percent of the genes in our genome are switched on and off at different times of the day (these estimates are still quite rough).[11] This is not surprising, because different tissues have different specialized functions for which they need different tools; thus, not all genes in our genome are used in all cells of our body. Why should a liver cell, for example, produce opsins, like the retinas of our eyes? There are, however, some genes that appear to go up and down in most cells of our body. Among them are those that encode the clock proteins, which can produce a circadian rhythm at the cellular level. This suggests that a circadian clock is potentially ticking away in every cell of our body. Clock researchers have known for decades that single cells can produce daily rhythms because they could observe them in single-cell organisms.

10

The tiny creature danced, together with millions of others, below the surface of the ocean. They swam toward the sun, many of them to the same location, forming dense clouds. When the clouds reached a certain density, they suddenly seemed to become heavier than water and sank away from the water's surface, glittering in the morning sun. The clouds dispersed into single creatures again, then immediately reformed and swam back up toward the sun. This repeating ritual created a tiny current, similar in shape to a magnetic field, drawing in other creatures at the surface. This accumulation made the clouds even bigger and denser, sending them down even farther. Currents of dancing creatures formed in many places, and neighboring currents started to move toward each other, the bigger ones drawing in the smaller ones. When the clouds were close enough, they fused into one, gradually forming long sheets, like streaming curtains or polar lights hanging from the ocean's surface.

This marine ballet continued throughout the day until, late in the afternoon, the clouds became thinner and the strength of the currents lessened. The little creatures seemed to have lost their urge to swim upward, and the swarming curtains began to dissolve and disappear. More and more creatures were now beginning to sink—endlessly, it must have seemed to them—until they reached the coastal bottom or a layer where waters of different temperatures met, forming an invisible blanket that was impenetrable to the tiny particles.[1] When the sun had set and darkness surrounded the large mats of creatures meeting at the bottom or at the invisible blanket, a spectacular display began. The swarm started to give off an ever-so-faint glow. When they bumped into each other, they emitted flashes of blue-green light that were a thousand times brighter than the faint background glow.

The tiny creatures were often hunted by others that were many thousand times larger than themselves. When these hunters swam into them they collided with thousands of their population mates, triggering a fireworks of flashes. Some hunters were startled, stopped, and forgot to eat, others turned around and swam away because they feared the fireworks would give them away, so that their own enemies would find and eat them.

The tiny creatures loved to swim toward light. At night, they made their own light in the otherwise pitch-dark depth of the ocean, and it seemed as if this self-made light at night helped them to stay together. If each of them swam toward that light, none would get lost. During the day, staying together was easy thanks to their gregarious dance, which herded them by their self-produced currents.

When winter came and the ocean's temperatures fell, the little creatures sank to the bottom, surrounded themselves with a hard coat, and settled in the mud, where they hibernated until the water temperatures rose again in spring. Once they detected the advent of warmer days, they got rid of their hard winter coats and got ready to mate. Throughout the year, their population grew by mere duplication of themselves. Having sex was their way of celebrating spring.

So far, we have been looking at the biological clock mainly in the context of our own existence, even if we have occasionally used mice and hamsters, or even bread mold and fruit flies, as models for our mammalian body clock. I hope to have convinced you in the last few chapters that the body clock is a profound biological function. It involves clock genes and their protein products giving it individual heritable qualities (for example, chronotype); and it controls the timing of our body on practically all levels, from switching genes on or off to changing our behavior. But how can we, who think of clocks predominantly as mechanical or electronic devices that help us keep appointments, understand the importance of this internal clock for survival?

To fully appreciate the importance of biological clocks and how they allow organisms to occupy niches they otherwise couldn't cope with, we do best to turn away from our own species. That is why I chose a quaint and anthropocentric way of telling a story about little ocean creatures. The protagonist is a marine alga, no more than a single cell armored by hard plates covered with little holes, as if it had been hit by a shotgun. If we would line up thirty of its kind, they would only measure one millimeter.

The alga's current scientific name is *Lingulodinium polyedrum.* I worked with this alga for more than fifteen years, first in Woody Hastings's lab at Harvard University and then in my own lab in Munich. At that time, it was called *Gonyaulax polyedra,* and had become a model organism for clock research because the single cells can produce light by a biochemical reaction called *bioluminescence,* which they do exclusively at night.[2] This bioluminescence could readily be used to record the biological clock ticking away in a single cell. The

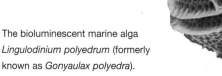

The bioluminescent marine alga
Lingulodinium polyedrum (formerly
known as *Gonyaulax polyedra*).

genus *Gonyaulax* belongs to the phylum of dinoflagellates, micro-scopic unicellular plants that can actively swim with the help of fla-gella, which they use like propellers.

Clock researchers have to look after their experiments at all times of the twenty-four-hour day. When I worked at Harvard, I entered the air-conditioned rooms where we cultured our *Gonies*—as we fondly called them—around the clock. I noticed that the cells swam near the surface when the lights were on but that they formed an in-active carpet on the bottom of their flasks in darkness, when they also displayed their self-made light. While *Gonyaulax*'s biolumines-cence was an established tool for studying the biological clock, no one seemed to be interested in the behavior of these creatures. I was fascinated by their drastic changes in behavior day after night after day. I was even more fascinated by the fact that these living dust par-ticles had any behavior at all and that their behavior appeared to control the entire population. I began to experiment with time-lapse recordings. What followed was a long line of research that offered many insights into the ecological significance of biological clocks and finally led to the amazing discovery that a single cell could con-tain more than one circadian clock.

The results gained from laboratory experiments, together with observations in nature, can give us insights into how *Gonyaulax* populations behave under the natural conditions of their environ-ment in the ocean. During the day, they rise to the surface to cap-ture light, which they need to produce energy and sugar molecules.[3] However, the upward urge turned out to have nothing to do with light.[4] If we offer the cells light from the side in the laboratory, they still swim upward during the day, showing that this part of their be-havior is guided purely by gravity. But under these artificial condi-tions, the cells also show a clear light orientation, even though it changes over the course of twenty-four hours. While the cells care-fully choose their position in a light gradient during the day (neither too bright nor too dim), they behave like moths at night: they will always swim toward light—even if it is so bright that they die. These experiments showed that the orientation behavior of *Gonyaulax* consists of two components—one using gravity *(gravitaxis)*, the other using light *(phototaxis)*. Both of these behavioral components are controlled by the cells' circadian timing system. The phototaxis is "choosy" during the day and mothlike during the night, and the gravitaxis is negative during the day (swimming upward) and positive during the night (sinking). The switch between swimming up and sinking is thrown two hours before the sun sets. But why do the cells sink to such considerable depths?[5] It could be exhausting to swim upward, but that is probably the easiest challenge for these little creatures. Although the surface offers a lot of energy in the form of light, the building blocks for making sugar, as well as other nutrients essential for their survival, are practically absent in surface waters.[6] Nutrients are released primarily by dead organisms, which are usually heavier than water and, therefore, sink to the bottom or are caught by a thermocline. That is why organisms that need light to photosynthesize but also other nutrients to proliferate face a dilemma. To get everything they need, they often engage in long, vertical journeys in the ocean; and to change their orientation at the right time, they need an internal clock.[7]

The "decision" to leave the surface toward the evening and wander off to get nutrients is an easy one because photosynthesis doesn't work in darkness. Deciding when to rise to the surface again while looking for nutrients in the depths of the ocean is more difficult. Let's assume that a swarm of *Gonyaulax* has harvested enough energy and sugar during the day at the surface. To proliferate, it needs the additional resources present only at lower depths—so, it sinks. But what happens if it doesn't find those vital other building blocks? Should the algae stay at depth until they eventually find the necessary goodies? Or should they rise to the surface because they are losing energy during every minute of their search? It is the circadian system of these cells that makes these difficult decisions. Depending on the internal time of day, nutrients such as nitrate will delay the internal clock and thereby prevent the cells from rising. But if a cell hasn't found nutrients all night and then, at last, encounters a source of them close to or even after dawn, lingering even longer at great depth could be lethal because the cell may run out of energy no matter how many nutrients it might assimilate. So, if nutrients are only found at the end of the night, they do not delay the internal time, and the circadian clock will throw the switch that lets the cells rise back to the surface. Only cells with a precise internal timing system can perform optimally in this game involving space, time, and nutrients, and only these can actually occupy this difficult ecological niche where resources are separated by long distances in addition to being available (as is the case with light) only at certain hours of the day.

To understand the complexity of circadian systems, it is advantageous to monitor more than one "hand" of the internal clock. Recall the results from human bunker experiments in which the activity–rest rhythm adopted a different period from the body temperature rhythm. The internal desynchronization of these two rhythms would not have been discovered if at least two different outputs of the circadian system had not been recorded. In the case of *Gonyaulax,* the clock was investigated for decades predominantly by recording bioluminescence. With the discovery of the behavior rhythm, it was possible to do long-term recordings of both rhythms simultaneously. Under certain light conditions in the laboratory, we noticed that the rhythms of bioluminescence and behavior showed *internal desynchrony*—like some humans did in the Andechs bunker. This result indicates that the circadian system can consist of more than one internal timer, whether in complex creatures like us or in single-cell organisms. In bioluminescent algae, one of them appears to be dedicated to optimizing the cell's responses to light, and the other seems to be responsible for coordinating the acquisition of nutrients. One of them is linked to photosynthesis and phototaxis, the other to nutrient uptake and vertical migration.

How advantageous it is to have an optimally tuned internal timing system was elegantly shown by Carl Johnson. He performed experiments with an even simpler organism than *Gonyaulax,* namely the blue-green alga *Synechococcus elongatus*.[8] These very simple single-cell organisms also have an excellent circadian clock. Like *Gonyaulax*, they get their energy via photosynthesis. Johnson took advantage of mutant strains that show very different free-running periods, one of them much shorter than twenty-four hours and the other much longer. When he cultured populations of these mutants in cycles of twelve hours of light and twelve hours of darkness in separate flasks, each one of them grew at about the same rate. He then mixed the two cultures in one flask so that the two genetically different strains were now competing for the same resources. This permitted him to ask whether one strain would outcompete the other. When he cultured them in constant light, both strains still grew with the same rate, neither outgrowing the other. However, when he grew them in a cycle of ten hours of light and ten hours of darkness (adding up to a twenty-hour day), the short-period mutant outgrew the long-period mutant. When he grew the mixture in a cycle of fourteen hours of light and fourteen hours of darkness (adding up to a twenty-eight-hour day), the long-period mutant won the growth race. Thus, when two strains compete for the same resources, the strain with an internal timing system that is most adapted to its temporal environment has the greatest advantage. Carl Johnson later also tested strains that had a dysfunctional circadian clock and found that they grew even slightly better than all other strains in constant light. In light–dark cycles, however, they didn't stand a chance against strains with a functional clock. These examples make it quite obvious how advantageous an internal timing system is for organisms that have to cope with a rhythmic environment in which resources are not constantly available.

You may have asked yourself why *Gonyaulax* cells glow at night. Scientists have thought of several advantages for bioluminescence, especially if the cells produce bright flashes when they are bumped into. Most marine biologists favor the "startle-your-enemy" hypothesis along with the "burglar alarm" hypothesis, presuming that the algae's enemies don't like it when they become literally visible to their own enemies by microscopic floodlights. I have an additional explanation that doesn't contradict the deterrent hypotheses. Many animals use bioluminescence for communicating with other members of their species. Think of fireflies flashing. For organisms that produce offspring by sexual reproduction, it is utterly important either to stay together as a population or to evolve some kind of communication to find others of their kind. *Gonyaulax* obviously belongs to the former and not the latter group. When they shed their winter coats *(cysts)* after having spent the cold season wrapped in ocean mud, they go through a transition for sexual reproduction. It is, therefore, imperative that they stay together—a lonesome Gony will not be able to exchange its DNA. The strange, mothlike affinity to light at night, when there is usually only the light they produce themselves, suggests that they may use bioluminescence to stay together.

The American clock researcher Mary Harrington was so taken by these algae and their temporal ecology that she wrote a poem about the clock-controlled life of *Gonyaulax polyedra*.[9] It is the only poem that has been published to date in the *Journal of Biological Rhythms* and perhaps the only poem ever published in an otherwise purely scientific journal.

> *FEEDBACK**
> If the lazy dinoflagellate
> should lay abed
> refuse to photosynthesize,
> realize:
> the clock will not slow
>
> but it will grow faint
>   weaker
>
>   weaker
> barely whispering at the end
>   "rise"
>
>   "rise"

to little effect.
The recalcitrant *Gonyaulax*
arms crossed
   snorts
"No longer will
they call my life
(my life!)
'just hands'.
I am sticking to the sea bed!"

* After reading Roenneberg T, Merrow M (1999) Circadian systems and metabolism. *J Biol Rhythms* 14:449–459.

11

Someone always tended the fire in the center of the great cave. The clan had some kind of arrangement as to whose turn it was to watch the fire, but these arrangements were only for emergencies—just in the rare case that nobody was awake. Normally, someone would be awake at all times during the night and would come to sit around the fire to chat or just add some wood. No one was able to sleep through the entire stretch of a night's darkness, except for the few short midsummer nights. Most members of the clan retired shortly after dusk: some earlier, some later—the youngest adults usually were the last to go.

Mrk, the clan's chief, had gone to his small, private side cave around dusk and had fallen asleep immediately. After some time he woke up, but not fully—entering a state between sleep and wakefulness. The clan's storyteller had once told him that this was the time when he developed his best tales. He claimed that, during these half-awake spells, he was able to see into worlds much stranger than their own. After lying in the dark half-conscious for a while, Mrk went back to a deep sleep, only to wake up later—this time fully. He got up and joined other clan members who sat around the fire in the big cave. The older Mrk got, the longer he found himself sitting around the fire. He had woken up during the night all his life but usually felt tired after a short while and spent the rest of the night sleeping on his mat.

As Mrk watched the dancing flames, his thoughts wandered off to the dark forest. While they were safe in the cave labyrinth, Urf, the man of his daughter and the current leader of the night-hunters, was out in the dark forest together with three others. Around sunrise, the night-hunters returned from their outing with their prey. The older members of the clan were already up and started to cut the large prey into smaller parts with handheld polished stone axes and flint knives while the hunters lay down to recover their lost sleep.

In this chapter, we return to my favorite animal—our own species. I promised in the introduction that the facts concerning the internal timing system underlying all stories in this book were sound. With regard to the preceding narrative, however, I hope you won't mind my taking a few creative liberties concerning the history of human life in the Stone Age. My aim is to take you back to our not-too-ancient history and try to imagine what life was like when humans were almost completely dependent on night and day, dark and light, cold and warm.

While writing about the Stone Age, I found myself using, and then rejecting, expressions like "several hours later." The concept of hours is of course inappropriate for that era, but it's an issue in our era, too. Not so long ago I was approached by a younger colleague, who inquired whether we had developed a different chronotype questionnaire for the "third world."[1] His girlfriend was researching human behavior in Madagascan villages that have no contact with modern civilization. Thus, asking inhabitants at what "time" they went to bed or got up was pretty useless. So far we haven't produced a chronotype questionnaire for these populations because it is difficult to exchange the useful time conventions of hours and minutes that we are so used to with another, equally measurable time frame. Sunrise, sunset, and possibly midday are natural markers for time of day. But in order to develop a finer time frame by using references from the environment, we would need to understand the daily routines of the population under investigation. Are there natural time markers, such as the appearance of birds, the opening and closing of flowers, shadows (cast by mountains), or specific odors during the night?[2]

How did our present-day culture of sleeping develop over the course of our evolution? Few animals actually sleep and are awake in a consolidated way over long stretches of time—many of them are *episodic* sleepers. Although they do sleep more at night than during the day (if they are day-active creatures), they tend to sleep whenever they have nothing to do (hunt, graze, or whatever). The description of humans gathering around the fire at night is not my ad hoc invention. Anthropologists and human ethologists reported this nocturnal behavior after returning from expeditions to populations not in contact with modern civilization.[3] Nights are usually much longer than we require for sleep. Near the equator, nights are about twelve hours long throughout the year. The farther we travel from the equator, the shorter the summer nights and the longer the winter ones. What did our ancestors do in winter when they had no light to fake a longer day? Even before the invention of the light bulb, light was commonly available at night. Night lights existed from the moment we were able to manage fire in a small container or in the form of a lit candle or torch. In a way, the central fire in the great cave was the beginning of light at night.

But even now that we have become completely independent of the night's darkness, we sleep and are awake for very long stretches of time—on average, eight and sixteen hours, respectively. We have already covered the main factors that determine when we fall asleep: internal timing and the amount of sleep pressure we have built up over the time we are awake. The largest decrease of this sleep pressure occurs during the first half of our sleep, so that we are relatively close to the wake-up threshold throughout the second half. We know from experience that the closer we are to our usual end of sleep when we wake up, the harder it is to fall asleep again. Charles A. Czeisler, a clock and sleep researcher at Harvard University, developed a theory about why we are able to sleep and be awake for such relatively long stretches of time. Alertness and sleepiness are controlled by centers in our brain, which obviously have to be able to record how long we have been awake. They also receive an input from the SCN to incorporate internal time. There is, however, another factor that has an impact on how sleepy we feel: our body temperature, specifically, the temperature of our brain.[4] Our wakefulness is highest in a narrow temperature range around 37°C. We get very sleepy when we have a fever, and reports of polar expeditions state that freezing to death resembles falling asleep. We feel tired when our brain cools down, and we feel awake when our brain warms up (if it doesn't become too hot). Czeisler's hypothesis is based on the fact that our body temperature reaches an all-day low during the second half of our sleep and thereby enables us to continue sleeping for another couple of hours despite most of our sleep pressure having been relieved.

This hypothesis could explain why we are able to sleep for such a long time, but we still need an explanation for why we can stay awake for an even longer stretch. While the first half of sleep is the most effective in reducing our sleep pressure, we build up most of our sleep pressure during the first half of our wake period, so that we are already quite sleepy around lunchtime—even if we don't eat a midday meal (if we do, the lunchtime dip is even deeper). Depending on how well and how long we slept during the preceding night, we can either fight this lunchtime dip or, if we cannot muster any resistance, doze off—at least for a power nap. Light helps to win the battle for wakefulness while darkness fights with great success on the other side. I have often witnessed colleagues dozing off into a deep sleep in the middle of an interesting lunchtime seminar, especially in former days when the seminar room had to be sufficiently dark to see the slides projected by an old-fashioned slide projector. Once we have survived this lunchtime dip, the likelihood that we will fall asleep declines again. Czeisler argues that this is because the body clock cranks up our body temperature again. According to his theory our consoli-

dated sleep and wake times are supported by the circadian temperature rhythm, making us colder in the second half of the night and
making us warmer in the second half of the day.

Mediterranean culture has developed the nap into a full-blown
siesta. The mechanisms behind this second daily sleep episode involve heat, darkness, and nocturnal sleep deprivation. In the Mediterranean summer, it is too hot for farmers to work outside for many
hours around midday. The amount of work per day is, however, the
same compared to that of farmers living in more temperate regions
who can work throughout the entire day—except perhaps on some
exceptional days during the summer. The quandary of the Mediterranean farmer was solved by starting to work very early in the morning and working until very late at night, but not around the hottest
hours of the day. Starting early and finishing late shortens sleep at
night, leading to considerable sleep pressure during the day. The
most natural reaction to the heat of the midday sun is to go into a
cool place, which was always a dark space before the use of air condi-
tioners. The combination of accumulated sleep pressure and darkness can easily induce sleep, especially if we lie down at a time when
most of us experience a lunchtime dip.

The siesta culture shows that there is some flexibility in how and
when we attend to our individual sleep needs. We can obviously split
our sleep into two major episodes, one at night and another at midday. Under special conditions, we can even consolidate our sleep into
one coherent episode of more than twelve hours, as some subjects in
the Andechs bunker did. But even under these strange conditions,
the average sleep duration amounts to somewhere between seven
and eight hours per twenty-four hours. Our flexibility in how and
when we catch up on our sleep need is, however, limited by our internal timing system, as Sergeant Stein discovered during his sleep
study. At some times we can fall asleep easily whereas at others we
find it almost impossible—no matter how exhausted we are. The
lunchtime dip is an excellent window to position a second episode of
sleep during the day. But remember, the local time at which an indi-

vidual experiences "lunchtime" depends on internal time—for extreme late types it might be as late as 6 P.M.

What we haven't addressed yet is the issue of how sleep patterns change with age. The older our fictional Mrk got, the longer he found himself sitting around the fire. Do these changes concern merely how we distribute our sleep within the day? Do our sleep needs change with age? Or are we even different chronotypes at different stages of our lives?

The End of Adolescence

12 Urf had been lying on the forest floor for a very long time but nothing had come along the trail so far. He had lost any sense for how long he had been lying there in almost complete darkness, surrounded only by the noises of the nocturnal forest. He and the other three hunters had left their dwellings at dusk and had walked for half of a summer's night before reaching the part of the woods where they hoped to find the deer. It was a good place—the deer had to pass through a narrow opening between two mountains to get from the dense and safe parts of the forest to the rich grasslands. They knew that they had reached their hunting ground too early—the deer wouldn't pass this spot until sometime before sunrise. It had taken them much longer to reach their destination than it would take them to go home to their dwellings. The shortest way between here and home led through the dense part of the woods, which was an excellent choice for their homeward journey once they were carrying their heavy prey. The shortcut was, however, a bad choice for the outward journey when they didn't want to announce their presence to all of the forest's animals—especially not to those they were trying to hunt. Today, the wind was blowing in the direction of the grassland. That was the best condition for hunting in these grounds: if the hunters took the precaution of making a wide detour around the big forest, the deer could not pick up their scent.

It was essential for the hunters not to fall asleep while lying in the dark for many hours on end, without moving or communicating. The forest—their potential prey and the potential predators for whom they might be an easy bounty—was to take no notice of them. They had walked fast, almost running, on their journey in the night, circumnavigating the forest for three-quarters of the way. But then they stopped and washed off their sweat in a small stream and rubbed themselves with moss and forest soil. They walked the last lap to their destination very slowly and then positioned themselves strategically around the narrow exit of the forest. Then they covered themselves with leaves, moss, and soil and embarked on their endless wait. They had learned all the tricks of night-hunting from their elders, whom they now had succeeded in the job. The long walks, the long waits, how to stay awake during the odd hours of the night, how to carry the heavy prey over long distances on as few shoulders as possible—all these were activities endurable only by men of a relatively narrow age range.

One way to stay alert was to keep the mind occupied. Urf thought of his woman and his children, who were safe back in the caves. He was one of the most successful members of the younger generation, and because of his success, he had been chosen by the chief's daughter. They already had two children and his woman was pregnant with a third. Urf and his family didn't sleep in the big cave, where the majority of the clan members had their resting places, but resided in one of the small side caves of the labyrinth. He remembered the many challenges he had had to face before he was recognized as one of their future leaders. When he was chosen to accompany the night-hunters as their apprentice, his teachers made him stay up several nights in a row and even made sure that he couldn't catch up on sleep on the day before he was allowed to accompany them on a real hunting trip. Now that he had been the leader of the night-hunters for several winters and summers, he and his fellows would have to choose a new young man to be taught all the tricks of the trade. After another couple of seasons, Urf would have to step down because he would not be able to lie quiet and motionless for most of the night without either falling asleep or becoming restless.

Different ages are associated with different sleep patterns. Newborn babies are governed by a feeding rhythm that is much shorter than twenty-four hours. Over the course of several months, their behavioral rhythms become longer until they gradually develop a daily pattern. This doesn't mean that the circadian clock doesn't function properly at that early age; all it means is that consolidating sleep and wakefulness into long stretches of time would be counterproductive since sleeping, feeding, growing, sleeping, feeding, and growing are the most important aims at that stage—in addition to shaping the development of the brain by sensory inputs during the short wake episodes. Human life is often depicted as a circle: we end up just as we arrived—with no teeth, wrapped in diapers, unable to sleep through the night. Although this may be true in many respects, the similarity of babyhood and old age is often only superficial. The reasons for not being able to sleep through the night and for falling asleep frequently during the day are very different at the two ends of our existence. Eus Van Someren, a Dutch sleep researcher, has shown that lack of light is a main cause for the *ultradian* sleep pattern in the elderly and especially in the mentally ill.[1] Most elderly people hardly ever get the chance to go outside, and the television is often their main light source. As a result, their body clock is not synchronized properly. Van Someren recorded the activity of subjects living in nursing homes and could thus quantify how much their nighttime sleep was disrupted by activity and, conversely, how much their daytime activity was disrupted by naps. He displayed these data as an activity profile, recorded with an actimeter.[2]

Although we can see clear differences in the amount of activity between night and day, a lot of activity was still recorded during the night (between 8 P.M. and 8 A.M.). These disruptions became much less frequent when his team installed bright light sources in the common rooms of these homes.

There is no doubt that our body and our brain change with age and that many functions work less efficiently. The centers in our brain that control rest and activity, sleep and wakefulness, are not exempt from these age-related changes. The experiments performed by Eus Van Someren show, however, that the age-related changes in our behavior are not necessarily a direct result of degeneration in the brain but are caused indirectly by other changes in our lives, such as lack of light and activity.

But what about chronotype and age? Does our internal time

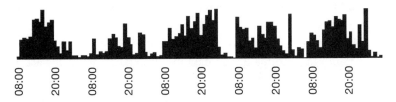

The nighttime sleep and daytime activity of patients in nursing homes are heavily disrupted. Redrawn from E. J. Van Someren, A. Kessler, M. Mirmiran, and D. F. Swaab (1997). Indirect bright light improves circadian rest–activity rhythm disturbances in demented patients. *Biological Psychiatry* 41:955–963, with permission from Elsevier.

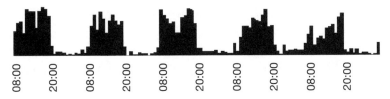

The introduction of bright light in the nursing-home common room alters activity patterns among patients. Redrawn from E. J. Van Someren, A. Kessler, M. Mirmiran, and D. F. Swaab (1997). Indirect bright light improves circadian rest–activity rhythm disturbances in demented patients. *Biological Psychiatry* 41:955–963, with permission from Elsevier.

change across our lifespan despite its genetic determination? Genetic control does not preclude change over time: our body height is, for example, genetically determined, but we don't reach that goal for many years, and we shrink when we grow old.[3] Similarly, chronotype changes with age—most noticeably in teenagers. Most of us experience twice in our lives (once when we are in our teens and once again when our children reach that age) the fact that teenagers can stay up easily until the early morning hours and possess an unchallenged ability to sleep through the day—almost the entire day. This may seem obvious, but it has to be proven with quantifiable data, and we want to understand the underlying causes and mechanisms. Epidemiology provides a good starting place for quantifying and eventually understanding a phenomenon—an approach that needs really big numbers.

A colleague once asked me whether it was scientifically ethical and sound to continue to add data to our database while the art of statistics had taught us to extrapolate from smaller samples to the entire population. She had a point, because it would be overkill to collect data beyond the point at which we can make good predictions with the help of statistics. This is certainly true if you only collect data to answer one specific question. The true value of a growing database is that you can constantly come up with new questions that go into more and more detail. If we wanted to know, for example, the chronotype distribution for every age group between ten and seventy (amounting to sixty-one age groups), separately for men and for women, then we would need 122 categories. If we wanted to have some statistical security for the results in each of these categories, we would like to have at least as many individuals within each group as we have groups. So in this case, we multiply 122 by itself, and the database should contain the information of at least 14,884 people. If we wanted to go into even finer detail, for example, by comparing those who live in the countryside with those who live in towns, the database ideally would have to be quadrupled (59,536). Our Munich ChronoType Questionnaire database is huge and steadily growing (at present more than 120,000 individuals, adding approximately 300 new entries per month), so that we are able to do what we just described. We looked at the changes in chronotype during development and across a human lifetime.

Young children are relatively early chronotypes (to the distress of many young parents), and then gradually become later. During puberty and adolescence humans become true night owls, and then around twenty years of age reach a turning point and become earlier again for the rest of their lives.[4] On average, women reach this turning point at nineteen and a half while men start to become earlier again at twenty-one. This one-and-a-half-year sex difference is typical for many aspects of human development. Although adolescence is often used to describe the stage between puberty and adulthood,

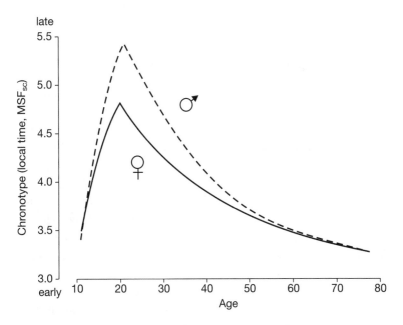

Chronotype depends on age. Teenagers belong to the latest chronotype in our populations. Women are generally slightly earlier chronotypes than men.

its end hasn't been clearly defined. When we saw this clear turning point in the developmental changes of chronotype, we realized that this was the first biological marker for the end of adolescence.[5]

Since teenage boys continue to delay their clocks for one-and-a-half years longer than teenage girls, men are on average later chronotypes than women during adulthood. This difference becomes smaller with age and vanishes at around the age of fifty-two. This coincides statistically with the age when women enter menopause. Of course, entering menopause isn't the reason for the disappearance of the sex difference in chronotype. The reasons for menopause are changes in the hormonal cocktail being produced at different ages. Men also go through substantial hormonal changes with increasing age. The decrease of testosterone levels, for example, is one of the reasons that men lose their six-pack abdomens and develop a belly. In a way, the graph of changing chronotype shows changes of hormones that start at the age of twenty and then gradually progress throughout our life until they result in the obvious features that initiate our old age, or, to put it in more biological terms, our post-reproductive stage.[6]

In my view, these developmental changes have clear biological roots. I often have had to defend this view against purely cultural explanations, which frequently come in the form of the *disco hypothesis:* if teenagers would only go to bed earlier (in other words, not party until the early morning hours), they would be able to wake up fresh as daisies to attend school. We have meanwhile collected statistics from many different parts of the world and found that the phenomenon of age-dependent chronotype is not exclusively a characteristic of modern and urban societies. It is found also in purely rural areas, from secluded valleys in the Italian Alps to rural Estonia, India, and New Zealand. The strongest argument against the disco hypothesis came, however, from experiments I learned about in two recent conferences. These experiments showed that similar changes in chronotype occur during the life cycle of rodents—lab rats that definitely never visited a disco.

If we are correct that a later bedtime during puberty or adolescence is a biological program and not simply a lifestyle choice, one could even turn the disco hypothesis around. If teenagers are so late in their internal time that they don't find sleep before the small hours of the night, then where can they hang out without disturbing the earlier chronotypes of society? Maybe discos are necessary niches or institutions for adolescents.

But why do adolescents have late chronotypes? In modern industrial societies, young adults around the age of twenty are (statistically) still involved in some kind of education and will not start a new generation (their own family) for several years. This is a relatively new state of affairs. In my story of the night-hunters, Urf is in his mid-twenties and has already been confronted with crucial challenges, the outcome of which have severely determined his future, including his place in the social structure of his clan. Urf obviously has "made it." The night-hunters, who provide the clan with high-quality protein by risking their lives, are regarded with great respect, as the leaders in their age group. Their qualities might have been similar to those of the heroes of our time, as for example the winners of Olympic medals. Qualities such as endurance, imperturbable focus and attention span, or physical excellence are limited to a relatively narrow age range, which roughly falls into the peak of lateness in chronotype. The average age of Olympic competitors at present is twenty-five, and it has been even younger in the past.

But why would being late be better than being early at that age? Is there more to being a late or early chronotype than just the timing of sleep? A possible answer to that question comes from highly artificial (and quite gruesome) experiments routinely performed by clock researchers. To investigate the control of the circadian clock over bodily functions, cognitive performance, behavior, or biochemical parameters without the interference of other factors, such as sleeping or running around, clock researchers invented "*constant routines.*" Well-paid and highly motivated subjects stay up thirty to forty-eight hours performing tests on an hourly basis. Unlike Sergeant Stein and his fellow soldiers, constant routine subjects are not allowed to sleep at all. Many physiological parameters, like blood pressure or body temperature, are controlled by the circadian clock, but also respond directly to changes in the environment or in our behavior. A change in posture, a big meal, moving around, even exposure to bright light may change the level of a physiological parameter so that its control by the circadian clock becomes masked. That is why constant routine subjects spend the entire time lying down in a dimly lit room, frequently eating small and isocaloric snacks while performing many different tests, giving blood or urine samples, or spitting saliva into a tube—all at hourly intervals.[7]

We and others have performed constant routines in which we assess the subject's chronotype with the Munich ChronoType Questionnaire before the experiment starts. There are several reasons for doing that. One is determine how much a short questionnaire can reveal about internal time: how well can chronotype, based on subjectively assessed sleep times, predict the circadian rhythm of other clock-controlled parameters? The other reason lies in the possibility of evaluating the data collected in a constant routine (or other circadian experiments), not only on the basis of local time but also on the more biological basis of individual, internal time. The results obtained for the validity of the MCTQ have been extremely encouraging. The daily ups and downs of biochemical factors, such as cortisol or melatonin, highly correlate with the midsleep (chronotype) assessed by the simple questionnaire. We have performed other validations of the MCTQ data, such as asking subjects to keep six-week-long sleep diaries and to wear actimeters. All these validations showed that people are remarkably precise in estimating their usual sleep times, and that an individual's chronotype relates to far more than merely his or her timing of sleep.

But how can a constant routine answer the central question of this chapter: what is the advantage of teenagers becoming late chronotypes? Constant routines are extremely challenging and, despite good will, certain subjects don't want to continue participating after the first eighteen hours or so. Some can be persuaded to continue and others plainly insist on stopping. Those subjects are almost always early chronotypes, who claim that they just cannot stay awake anymore. Logically the same thing should happen to late chronotypes a couple of hours later, but astonishingly it doesn't. It seems that an additional quality of being a late or an early type is related to sleep pressure. Early types apparently build up sleep pressure more rapidly than late types. In addition, early types don't have the ability, as late chronotypes do, to "sleep in" after having gone through sleep deprivation.

These insights into chronotype are relatively new, and therefore have to be treated with caution. Sleep timing and the speed with which an individual builds up sleep pressure may not always be tied to one another. There are certainly early types that can go on for a long time without sleeping and still perform with admirable focus and quality, but this ability seems to be much more prevalent in late than in early chronotypes. But if there is something to the relationship of being a late type and being able to endure and "sleep in," then late types are better night-hunters than early chronotypes. The lateness of teenagers in our modern industrial age may be a remnant of a skill that accompanies the age of peak physical condition.

But what does that mean for all those young people who have to perform in school during the early hours of the day? Note that in some regions of Europe, especially in Germany, schools start at seven o'clock in the morning—which means that some pupils have to get up at 5 A.M. or earlier in order to be in class on time.

13

Jacob needed a smoke and persuaded Felix to leave campus with him during the first morning break. He was in a really foul mood. He had slept far too little, his parents were on his back about his current performance in school, and now that thing in class again. His mother had talked to the teachers of his worst subjects. As a reaction—probably well meant by the teachers—they suddenly generated frequent opportunities for him to get higher oral grades to compensate for the bad ones he was getting in most of the written exams.

It had been the third time this week that Jacob had suddenly been confronted with a teacher's anticipating and friendly face radiating something like: *now Jacob, you can surely answer this extremely easy question.* This time it was their math teacher, Mr. Blossop. The trouble was that he had no clue what the extremely easy question might have been. "What *is* this, Felix?" Jacob almost shouted at his friend after lighting up, "I sit next to you in class, I listen, and then suddenly, out of the blue, he's in my face, and I don't even remember what he was talking about. It's like I had a complete blackout. Maybe Blossop was trying to help, but I just didn't hear the question." Jacob looked at Felix with desperation. "What did he ask me?"

"It was really simple," answered Felix. "All he wanted to know was what the first derivative of a function tells you about the qualities of the original function."

"That would be the slopes at every point of the original function," answered Jacob.

"Exactly, so why didn't you say so? You just sat there staring at him with a completely blank face as if he had asked you to prove it mathematically for all possible conditions."

"Because I hadn't even heard the stupid question! I already told you that! Are you having a blackout now?"

"I heard you, but Mr. Blossop doesn't know that, and this isn't the first time you've drawn a complete blank this week."

Jacob took a last draw of his cigarette and chucked the butt into the bushes, and the two of them headed back to school. "It's so unfair. Hilda just sits there in the first row, fresh as a daisy every morning, listening to whatever Blossop says with her big blue eyes wide open—catching every word of it. I bet she never has blackouts!" He looked at Felix, soliciting some support. "The first couple hours in school are a complete waste of time!"

Felix's answer came as a surprise: "I'm sure that's what some teachers think, too, when they have to deal with students like you." Jacob wasn't offended; he knew Felix far too well for that. First, Felix was much more alert than Jacob in the morning, and then he had this tendency to always see the other side.

Jacob thought that he was much better at math than Hilda but was never able to prove it. Why did they have to take all their tests first thing in the morning? He would have performed so much better at some later time of the day. Over the past few months, when he had realized that his performance in school was on the edge, he had really tried hard to go to sleep earlier on school nights. He had said "no" to every party invitation—even on weekends. But nothing helped—he just couldn't get more sleep on schooldays. He caught up on schoolwork during the late evening, sometimes until the small hours of the morning. It didn't help to go to bed earlier than 1 A.M., because he just couldn't fall asleep.

While they walked through the school's big hall toward the physics classrooms, Jacob noticed Ann, who lived on the same street as he did. Although she was quite a bit younger, he had thought about asking her out. When she turned around, Jacob read the slogan on the back of her tee shirt and smiled for the first time that morning.

A couple of years ago, I was invited to attend a hearing in the parliament of Saxony in Dresden. The opposition party had organized a debate on moving the beginning of school to a later time. In my short lecture, I gave an introduction into the nature and the mechanism of biological clocks and about their genetic background. At the end of my presentation, I described our findings on the peak of lateness in teenagers. Immediately after the chair of the hearing had opened the discussion, several schoolteachers—also members of the hearing panel—raised their hands. One of them, a physics teacher, seemed quite agitated and declared with strong conviction: "My students are fully awake at seven in the morning." I asked him what age group he taught, and he answered: seventeen to eighteen. I then asked this confident physics teacher what evidence he had for his statement, and he declared with no less conviction, "I can see that—it's perfectly obvious!"

It is quite remarkable how firmly belief and conviction stand in the way of reasoning. This was not a teacher of religion but of physics, who should understand the rules of science. What grade would he have given Jacob, for example, if he had answered the question "Is there evidence that the sun turns around the earth?" with a simple "Yes, there is—we can see it—it's perfectly obvious!" But ever since Galileo's time, scientists have agreed that only facts count in a scientific argument. After statements like these, my long-term collaborator, Martha Merrow, would frequently exclaim, "Show me the data!"

The next contribution to the discussion came from another teacher who was also the principal of a large school in a rural part of Saxony. He described in great detail the bussing system in his region and made it quite clear that the issue of biological clocks in youngsters was completely negligible. Both schoolchildren and workers had to be transported by the same bus company, and those buses had to be available for the workers at a certain time. Consequently there was no way to change the time of day when children could be bussed to school! That's it; very simple; end of discussion!

As you can see, the level of this discussion was frighteningly low and shallow. I have had conversations of similar quality with many other teachers and politicians throughout Europe. One of them was a remarkable exchange with Bavarian politicians, carried out with the help of a large Munich tabloid. The tabloid had picked up on our findings regarding the biological lateness of adolescents and wrote a provocative piece advocating later school times. The next day's issue of that newspaper contained an article under the headline "Sleep researchers caught sleeping." The content of the article was an interview with the spokesperson for the Bavarian Ministry of Education, who gave a very entertaining, seemingly logical counterargument proving the ignorance of clock and sleep researchers. Hadn't it been those very scientists who had found the infamous lunchtime dip? If school times were delayed—as those researchers obviously demanded—wouldn't the poor children then have to be taught during this time of tiredness and short attention span? How could these researchers be so shortsighted?

When someone brought an issue of that paper into the lab the next day, I immediately called the ministry and got hold of the spokesperson who had been interviewed for the article. I told her that I was extremely interested in the sources that prompted her to make these statements because we certainly must have overlooked an important paper, and we wanted to amend that oversight immediately. She was very friendly—as spokespeople probably should be, given their job description—and promised to send me the sources the next day. Let me specify the source that I waited for: a study showing that delaying school start times led students to be more tired at the end of a full morning's teaching than if school had started earlier. In many more words I was quoting Martha: "Show me the data!" The promised sources were never produced, not even after several reminders.

I could fill this book with similar debates, which would be highly repetitive and boring, but I will recount a last example to round things up. A German newspaper once invited me to participate in a debate with the president of the German Teachers Association. In no more than 200 words—without knowing what the other would write—we were to make our statements for or against the introduction of later school times. The president summarized his counterargument in four points. First, he asserted that the daily peaks of performance are individually too different to warrant later school times, so that times between 8 A.M. to 1 P.M. constitute an excellent compromise for all chronotypes. The president added that the reason for students not being awake in the morning was only—if we were really honest!—because they regularly went to bed too late. Second, he was convinced that the majority of parents wanted schools to start at 8:00 because they themselves had to go off to work. Third, he described at length the bussing difficulties, as had his colleague in Dresden. Finally, after talking to students, he had become convinced that 90 percent of them favored being in school by 8 A.M. because they wanted to be home by 1 P.M.

An astonishing set of statements! Again the arguments are predominantly based on conviction (a word my adversary used frequently throughout his text) and belief rather than on facts. Some of his arguments merely show a lack of information. It is delectable that the first statement acknowledges the existence of individual differences in chronotype. The argument overlooks, however, that the distribution of chronotypes within a given age group moves to later times during adolescence—everyone becomes considerably later, regardless of individuality. It also neglects the fact that the entire distribution—including young and old—is already so late in our present day and age that well over 60 percent of the population would have difficulty fully concentrating during the first hours of school. The statement about the traditional school times being an excellent compromise is, therefore, not supported by data. The first statement ends with the disco hypothesis. The tragic thing is that every one of the sub-statements is true: adolescents aren't fit in the morning—because they haven't slept enough—because they went to bed too late. All true—but why?

The president's second statement about parents being afraid of losing control over their children in the morning reflects a common misunderstanding. The discussion about later school times predominantly concerns adolescents. Why shouldn't high school students leave the house after their parents? His third statement about the buses obviously speaks to a difficult logistical problem—why else would it pop up so often when school times are discussed? Education is a nation's investment in its future. If there is a chance to increase the quality of such an important investment, then logistical problems, such as bussing, have to be solved on the long view—even if this is difficult. They should not be used as a valid argument against improving the investment in our next generation. Finally, the president presented no evidence for his fourth statement. With how many students had he spoken, and what were the questions? Might the students have answered differently if the question had been asked by a classmate? In addition, although I have witnessed innumerable changes in school policies, none of them would have been stopped just because students didn't want it. So after hearing a lot of beliefs and convictions, let's try and find some facts.

The most important fact has already been presented: the adolescent clock delays by several hours, reaching a peak in lateness at around twenty. The facts supporting a biological rather than a social reason for this delay are overwhelming. Mary Carskadon was the first scientist who recognized the biological rather than social basis for lateness in teenagers.[1] Her first papers showing this tendency were published in the early 1990s. Since then, many studies have monitored what happens to students when school times are changed. The results of these studies are extraordinarily clear. When Carskadon brings students into a sleep laboratory instead of sending them to school at their normal early time, she finds that many students show the signs of a major sleep disorder—narcolepsy.[2] When given the chance, they fall asleep at once and immediately enter a sleep stage called REM.[3] Normally this stage is more typical of the end of sleep than of its onset, showing that these students are physiologically still asleep, despite having gotten up in the morning. In their immediate entry into REM sleep, these students resemble patients suffering from narcolepsy. If people are so tired that they show a behavior normally found only in narcoleptics, they also tend to have increased episodes of micro-sleep. As a result of these episodes, people experience a short loss of consciousness, similar to what Jacob experienced. Often neither the affected individual nor people witnessing such events are aware of these micro blackouts. As a matter of fact, many car accidents that happen at night on empty roads are a consequence of micro-sleep episodes. The lucky drivers who survive such accidents have no recollection how they ran off the road.

Teenagers need around eight to ten hours of sleep but get much less during their workweek. A recent study found that when the starting time of high school is delayed by an hour, the percentage of students who get at least eight hours of sleep per night jumps from 35.7 percent to 50 percent.[4] The adolescent students' attendance rate, their performance, their motivation, even their eating habits all improve significantly if school times are delayed.

The president's claim that the traditional school times are a good compromise for all students isn't only factually incorrect. The traditional school times blatantly discriminate against late chronotypes, who make up the majority among teenagers. This isn't a good compromise for people who are not awake at seven or eight. Early types, however, could still perform just as well if schools started later. A German study assessed the chronotypes of university students and compared them with their grades on their final high school exams.[5] The resulting correlation is frightening. The later the chronotype of a student, the worse the grades.

The resistance of teachers and politicians to start schools later for teenagers is even greater in those systems where school is predominantly restricted to the morning, and teachers are accustomed to spending the afternoon at home correcting homework and preparing lessons. Delaying school times would certainly mean an adjustment to their schedules. As adults, most teachers have shifted to an earlier chronotype. There even may be a tendency in the teaching profession to self-select for early chronotypes. Despite these barriers, the number of schools trying out other timetables for adolescent students is rapidly increasing in several countries, from Switzerland to the United States.

A recent Danish project has eliminated timetables entirely and left the decision about when to arrive at school to the students.[6] One of the teachers of this school in Copenhagen recently pointed out in a television interview that schools should be regarded as service centers, and so they are required to offer the best possible service to their customers, meaning the optimal environment for achieving the best education possible. Allowing students to sleep and work at their optimal times should definitely be part of this service. Scientists are monitoring this school project, and I am eager to see the first results. But how far should we go in letting adolescents choose their wake and sleep times? Wouldn't that lead into a day-night inversion, so that teachers would have to start their classes at eight in the evening? To answer this question, we will explore in more detail the mechanisms (called *entrainment*) of how the clock synchronizes to the twenty-four-hour day.

14

It is the year 2210. The world never really recovered from the big economic depression two hundred years ago. Concern for the environment and nature goes down the drain in times of financial hardship. As a consequence, many regions of our globe have become unlivable for most plants, animals, and humans. The area usable for human settlements has been reduced to a tenth of what was available around the turn of the millennium and continues to decrease rapidly. Places to live are not the main problem—humans have retreated to higher regions, shielded from the rising ocean levels, or live in fantastic underwater worlds. Area to produce the necessary resources to support Earth's remaining human population is vanishing, however: agricultural land hardly exists and the formerly food-rich oceans are overfished and overpolluted.

At the headquarters of the World Agency for Space Settlements, WASPS, which has recently been moved to the summit of Mont Blanc, a committee of scientists meets to discuss the possibilities of evacuating their troubled planet. The exploration of other planets in our Solar System has been quite successful, and human settlements have been established on most of them. Although environmental issues have been neglected, research, technology, and development—RTD—has reigned supreme. The technical problems involved in creating an earthlike environment have been solved on most planets. Scientists know that it is essential—in the long run—to create settlements that are somewhat in contact with whatever one would call nature on these ersatz planets. They decided to build the settlements under huge, transparent domes so that the inhabitants had at least some access to natural light, something that could be called weather, and—at night—to twinkling stars.

The committee had invited a chronobiologist to today's WASPS meeting. She had written to them to point out that the agency had overlooked a serious problem in their evacuation plans. The problem lay in the different lengths of days that every planet produced. Human clocks, she had noted, are quite fussy about day length, and malentrained clocks cause huge health problems, as we know from the early days of shift work. The concerned chronobiologist, Svenja Rasmunson from the Center for Chronobiology in Tromsø, was asked to give a more detailed presentation about her concerns and their scientific background to the committee. During her talk, the committee members looked at their individual screens on the long conference table. One image showed a list of planet rotation times, corresponding to the length of the planet's day in hours:

Mercury: 1392.94
Venus: 5771.25
Earth: 23.93
Mars: 24.62
Jupiter: 9.92
Saturn: 10.23
Uranus: 17.73
Neptune: 18.20
Pluto: 151.50

Professor Rasmunson indicated that the hugely different day lengths on the other planets created a problem for the synchronization of the human clock. She proposed to determine the chronotype of every human on Earth, which could easily be done by loading a drop of blood onto a device called the "ChronoChip." With that knowledge one could assemble groups of similar chronotypes, and then send each of these off to a different planet. She then went into a long explanation of the formal basis of synchronizing circadian clocks—*the principles of entrainment,* as she called them.

Professor Rasmunson was interrupted by one of the other scientists. "What happens if partners or family members have very different chronotypes?"

"That shouldn't be a very frequent scenario," she responded. "As you know, human courtship behavior over the past 200 years has favored partnerships among similar chronotypes, thereby producing rather homogeneous families, similar to the historical family traits of social rank, ethnicity, or religion. If it were to happen occasionally, however, it might create a real problem." She then added, "We can do a fair bit with light intensities and wearing sunglasses to adjust individual phase." After a brief moment of contemplation, the professor added with a little chuckle, "On the other hand, some people might be quite pleased to have an excuse for a separation." She had recently been cheated on by her husband and was in the middle of a divorce, and her remark got a couple of laughs out of the audience.

The majority of people would surely be settled on Mars, which already had the biggest extraterrestrial colony in the Solar System. "According to the old proverb, every one of them will be healthier, wealthier, and wiser," quipped Svenja—but since only chronobiologists could easily appreciate the joke, nobody laughed. She quickly moved to the subject of special treatment for extreme chronotypes, like those who were first found in Utah. Individuals of this chronotype could potentially be sent to Uranus and Neptune, or maybe even to Jupiter and Saturn.

Several years later, WASPS decided to send everyone to the large Mars settlement and discarded all plans for settling on other planets. In their initial advertising campaign, WASPS even picked up the joke Rasmunson had made that day on Mont Blanc:

Off to the stars,
settle on Mars,
see good old Earth rise,
be healthy, wealthy, and wise!
(Discounts and prime housing opportunities for early risers–
fast deciders.)

This may be the most difficult of the twenty-four chapters for some readers to digest. But please bear with me and take your time. It will be worth your while because you will only really understand how body clocks synchronize to the twenty-four-hour day on Earth if you are familiar with the formalisms underlying synchronization, called *principles of entrainment*. Once you have understood these principles, you will easily understand why people with different clocks have to be different chronotypes, why your chronotype depends on whether you are an office worker or a farmer, and why Svenja wants to send extreme early chronotypes to Uranus or Neptune. You may wonder why I constructed a story around life on other planets when I want to explain how entrainment works on Earth. The reason is simple: you start to understand everyday life if you step out of it.

Few people know that the rotation of our planet has slowed down in its unimaginably long life history. This process has not stopped, so days on Earth will be close to twenty-five hours in a very, very distant future. Thus biological clocks can obviously adapt to different day lengths. This long-term adaptation is, however, different from sending people to other planets where they would arrive within months or years. The evolutionary adaptation generates different versions of clocks with altered or even new genetic components whereas the synchronization to days on other planets must be achieved via the principles of entrainment.

Successful entrainment ensures that the body clock produces

internal days that have on average the same length as the external days of the cyclic environment. You already know that the clocks of different individuals run differently (by inheritance) and may, therefore, produce days that are longer or shorter than the rotation of our planet (as if these people were made for a life on other planets). The principles of entrainment have to accommodate the following conditions in our extraplanetary scenario:

1. If individuals' clocks, producing internal days that are shorter than twenty-four hours, have to entrain to the rotation of Earth, then their internal days have to be lengthened.

2. If individuals' clocks, producing internal days that are exactly twenty-four hours, have to entrain to the rotation of Mars, their internal days also have to be lengthened.

3. If individuals' clocks, producing internal days that are longer than twenty-four hours, have to entrain to the rotation of Earth, their internal days have to be shortened (the most common scenario for humans).

4. If individuals' clocks have to entrain to the rotation of Neptune, then their internal days would *really* have to be shortened, even if they were producing internal days that were already shorter than twenty-four hours.

The idea of entraining the clocks of humans to the rotation of other planets is not far-fetched: Claude Gronfier simulated light–dark cycles longer than twenty-four hours in the laboratory to find out whether the human clock could entrain to days on Mars.[1] Humans can easily entrain to a twenty-four-hour day in the laboratory, even if the intensity of the light–dark cycle is relatively dim. Gronfier found that he had to increase this intensity to successfully entrain his subjects to the longer, Mars-like days.

But before I come to the question of how we possibly could entrain to days on other planets, I want to address a more general question: how does a circadian clock entrain to a *zeitgeber,* or external cue?[2] Technically, anything that produces a rhythm is a kind of oscillator (like a swing or a pendulum). But how can the period of an oscillator be adjusted to the period of another rhythm? Let's presume that the body clock is a swing that is being pushed by a zeitgeber (for example, the light–dark cycle). To synchronize the body clock, the zeitgeber has to systematically interfere with the momentum of the swing. (Admittedly, a swing is a strange representation of the body clock and the pusher an even stranger impersonation of a zeitgeber.)

At 6 A.M. the swing starts to move from left to right. It passes its lowest point at noon, reaches the other end by 6 P.M., and passes the low point again at midnight on its way back. In short, it swings from the left to the right during the day and back from the right to the left during the night. Whenever the zeitgeber pushes, the swing's reaction will depend on when during the swing this happens. If the swing

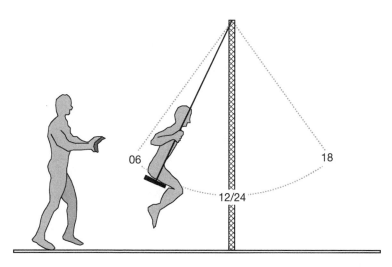

The motion of the swing represents the body's internal clock, and the pusher is the zeitgeber, keeping the swing in a precise twenty-four-hour rhythm. From T. Roenneberg, S. Daan, and M. Merrow (2003). The art of entrainment. *Journal of Biological Rhythms* 18(3):183–194.

is pushed while it is moving away from the pusher, it will speed up; if it is pushed while moving toward the pusher, it will slow down or even stop.

If a body clock could think and talk, it might respond thus to the following situations:

> If the clock thinks that it is past midnight (internal time) but suddenly sees light, it might say, "Oh, gosh, it's already approaching dawn. I really have to hurry up!"
>
> If the clock thinks it is already past sunset but then sees light, it might say, "Golly, it's still light out there, I really have to slow down and get back on track!"
>
> If the clock is notified by the eyes that is broad daylight *and* it thinks it is midday, it might lose its patience and respond, "I know it's day! Please bother me *only* if there is something to report that I don't already know!"

It is quite remarkable how the body clock responds to light. The experimenters who investigated how the clock responds to light used single light pulses, each of them given at a different internal time in otherwise constant darkness (or dim light). For each of these experiments they measured how much the clock changed the course of its internal time in response to the light stimulus. These experiments have taught us a lot about how the circadian oscillator responds to light, but their experimental conditions are highly artificial. In nature, circadian clocks are not completely dark-adapted over many days and are certainly not synchronized by a daily, single short pulse of light. We have, therefore, recently extended the theory of entrainment, originally constructed with single light pulses, so that it can explain entrainment in the "noisy" light environments of the real world. This new theory is based on a "response characteristic" that shows how sensitive the clock is toward light at different times during its internal day and indicates whether light shortens (compresses) or lengthens (expands) the internal day.

Entrainment is nothing more than making the internal day fit the external day—either by compression or by expansion. The response characteristic shows that light around internal dawn compresses the internal day (the higher the curve, the greater the compression); it has little or no effect around internal midday (the curve runs parallel to the dotted zero-line); and light expands the length of the internal day around internal dusk (the lower the curve, the greater the expansion). These specific responses allow biological clocks to synchronize with their cyclic environment. To correct for errors that may arise from the internal day being too short or too long, the body clock exposes different parts of its internal day to light and "hides" the remaining parts in the dark.

How do the rules of this hide-and-seek game explain the numerous cryptic aspects of entrainment on other planets? What is the relationship between chronotype and the day length, Svenja Rasmunson's main concern? The clock's free-running period (its internal day

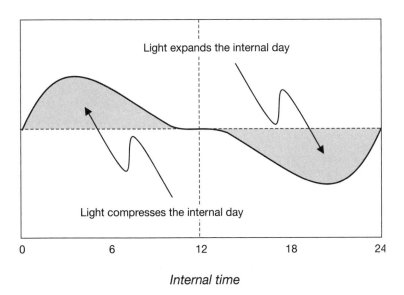

Internal time

Light entrains the internal day to the external day. Light around internal dawn compresses the internal day, has little or no effect around internal midday, and expands the length of the internal day around internal dusk.

length) is one of the reasons why we are different chronotypes. When early and late types live in temporal isolation, the clocks of the former run faster than those of the latter, producing shorter or longer internal days. If an internal day is shorter than the external day, it has to be expanded. If it is longer, it has to be compressed. What could be easier?

Let's start with the most simple of all entrainment cases, an individual's clock that produces on average internal days that are exactly twenty-four hours long. When it has to entrain to the twenty-four-hour light–dark cycle of our planet, nothing has to be changed; the clock's internal days should be neither compressed nor expanded. On very short (winter) days, this could be achieved by "hiding" all light-responsive portions of the cycle in the long nights. However, if

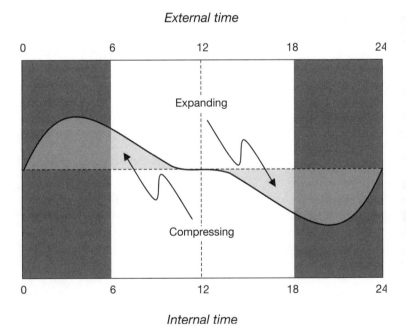

An internal clock that produces exactly twenty-four-hour days entrains by exposing equal amounts of the compression and expansion portions to the light, so that the two opposite effects cancel each other out.

days are longer (as in spring or summer), the sensitive portions of the response characteristic will be exposed to light. In this case, the only way that a clock which produces exactly twenty-four-hour days can entrain is by exposing equal amounts of the compression and expansion portions to the light, so that the two opposite effects cancel each other out. Note that these are the only circumstances in which internal and external clock time are identical (the two time scales are indicated at the bottom and the top, respectively). Internal midnight matches the middle of the dark period and internal midday occurs when the sun reaches its highest point.

Yet humans whose clocks produce exactly twenty-four-hour days are extremely rare—most human clocks produce longer days, which therefore have to be compressed. To achieve this, the clock simply

External time

Internal time

An internal clock with a longer day has to move to a later external time (see white arrow), thereby exposing more of the compression portion to the light and hiding more of the expansion portion in the dark. This results in a late chronotype.

exposes more of its compression portion to the light and "hides" more of its expansion portion in the dark. As a consequence, internal time will end up being a bit later than external time. The longer a clock's internal day, the later it will "move" in relation to external time. This is why people with slow clocks have to be late chronotypes; otherwise their clocks would not match the day length on our planet.

The opposite is true for clocks that produce internal days shorter than twenty-four hours. They have to move to an earlier external time, thereby exposing more of the expansion portion to the light and hiding more of the compression portion in the dark. Faster

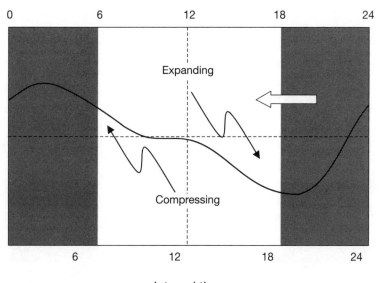

An internal clock with a shorter day has to move to an earlier external time (see white arrow), thereby exposing more of the expansion portion to the light and hiding more of the compression portion in the dark. This results in an early chronotype.

clocks produce earlier chronotypes, so that their internal noon is now earlier than external midday.

These two theoretical examples refer to the entrainment of individuals whose clocks produce different internal days but who all live on Earth. Since it matters only for entrainment that a clock's day has to be expanded or compressed, it is but a small step to understand entrainment on other planets.[3] If WASPS sends people to a planet that rotates faster than Earth, their internal days would have to be compressed (by moving internal time to a later external time), thereby creating later chronotypes. If WASPS sent the same people to a planet that rotates slower than Earth, their internal day would have to be expanded (by moving internal time to an earlier external time), thereby creating earlier chronotypes. Since the days on Mars are longer than the days on Earth, all of us would become earlier chronotypes, as Svenja's joke and the WASPS advertisement suggest.

But can our body clock adjust to the day length of any old planet? Surely not: how could we fit our activity times and our sleep needs into days that are approximately sixty times longer than ours (for example, on Mercury)? We wouldn't be able to function without sleep for forty days, and then we wouldn't be able to stay asleep for the following twenty days. So what are the limits within which our body clock can adapt to the days on other planets? The hide-and-seek game of entrainment makes the answer easy. The shortest day length we can entrain to is reached when the entire compression portion is exposed to light (and its entire expansion portion is hidden in the dark). And the longest day we can entrain to is reached when the entire expansion portion is exposed to light (and its entire compression portion is hidden in the dark). Let's assume that the body clock could be compressed and expanded maximally by two hours and that its internal day was exactly twenty-four hours long. Theoretically, this clock could entrain only to days longer than twenty-two hours and shorter than twenty-six hours.[4] The maximum compression and expansion capacities depend on zeitgeber strength: the

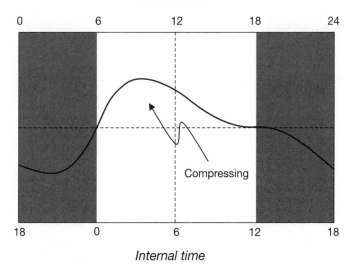

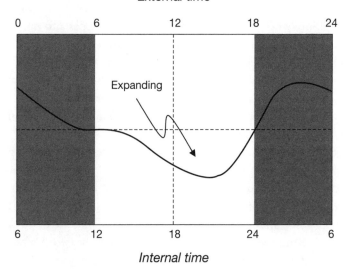

The limits of entrainment: the shortest day we can entrain to is reached when the entire compression portion is exposed to light *(above)*; the longest when the entire expansion portion is exposed to light *(below)*.

brighter the day and the darker the night, the wider the range of entrainment. Remember, Claude Gronfier had to use brighter light in his laboratory experiments to entrain subjects to a Mars-like day than was necessary to entrain them to a twenty-four-hour day.

According to these principles of entrainment, we have to presume that the internal days of extreme early chronotypes are very short.[5] That is why they are probably the only humans who might be able to entrain to the natural days on Uranus and on Neptune (with day lengths of 17.73 and 18.2 hours, respectively). In this highly theoretical scenario, these extreme early birds on Earth would become extremely late chronotypes on Uranus and Neptune, much later than the latest teenagers on our planet. Even so, it is unlikely that the internal days of these extreme early chronotypes would be as short as eighteen hours, making their outing to Neptune nothing more than a chronobiological fantasy. And yet, some mutations do have drastic effects on the length of internal days.

I now have covered all of the entrainment issues hidden in my tale of life on other planets except for one: why did Svenja Rasmunson suggest at the end of her presentation that extreme early types could even be sent to Jupiter and Saturn, which are characterized by day lengths of 9.92 and 10.23 hours, respectively? Surely even extreme early types cannot get up every nine to ten hours, be awake for six hours, and sleep for only three hours. This regime would resemble what Sergeant Stein had to live through—without much success, as we know. But interestingly, this wouldn't happen. The body clock would adapt to these extremely fast cycles by what chronobiologists call *frequency demultiplication*. You know frequency demultiplication well from everyday life situations. When people clap to music in concerts, they often clap on every other beat. If you walk swiftly with a stick or an umbrella, you poke it onto the ground every other step. On Jupiter or Saturn, the clock of extreme early types would simply skip every other day as if the second dark period was the result of a regular thunderstorm in the afternoon. Inhabitants would probably adopt a Mediterranean lifestyle on Jupiter or Saturn: they would sleep-deprive themselves during one of the (short) nights and have a nice long siesta during the other. Yet only extreme larks could live such a life on Jupiter or Saturn because "normal" clocks would fail to entrain to the eighteen- to twenty-hour double days on these planets.

15

It was 7 P.M. when Oscar, a middle-aged and slightly rotund surgeon, handed his new acquaintance two glasses: one with gin, tonic, lime, and ice; the other with plain tonic water. "I didn't know how strong you want your G&T, so I brought some extra tonic," he said. "I'm Oscar, by the way."

"Thanks for the drink, Oscar. My name is Jerry."

When Oscar had settled in one of the comfortable, overstuffed leather armchairs, he raised his glass, which appeared to contain a Bloody Mary. "Cheers." After taking a sip, he put the glass down on the stylish glass table in front of him and took a single peanut out of a small bowl. "Have you reached your wife?" he asked, continuing the conversation.

"She won't pick up the phone—we had a fight, and I told her to go to the dinner party on her own," answered Jerry, the younger and definitely fitter one of the pair. His face expressed both disappointment and anger.

Oscar looked at him with understanding. "I couldn't help overhearing that you were having a difficult time on the phone," said Oscar. "I'm sorry," he added after a little pause. "When I was still married, I found myself a victim of many similar quarrels. They just don't know what we go through in our jobs."

Jerry nodded slowly. He looked like he was about to fall asleep, and his gaze was turned inward. With a jerk of his head he forced himself back to reality. "Sorry, what did you just say?"

Oscar slightly rephrased his last statement. "Our spouses, they don't appreciate the constant stress we are under." This time Jerry's acknowledging nod was more engaging.

"What do you do for a living?" he asked. "Or are you already retired?"

"No, not yet," said Oscar. "I am a surgeon specializing in organ transplants." Unexpectedly the younger man broke into an almost hysterical laughter. A group of people sitting nearby turned to look at him. Oscar also looked at him in surprise. "What's so funny about that?"

"Oh, I'm sorry," murmured Jerry, still fighting his giggling fit. "It's just that I'm also an organ specialist."

"How nice, we're colleagues! Where are you based? Let me guess —Cincinnati? They are building up a new unit for organ transplantations there, and I had heard that some colleagues were also in town. We must have met before, perhaps at a conference. I'm so sorry— I have a bad memory for faces—it's names for me. What is your last name?"

Jerry's laughter became louder again. "I don't think that we have any reason to have met before—my organ is bigger than yours." Oscar thought that he had definitely given him too much gin. He didn't like where this was going—he loathed men's jokes. When Jerry saw the expression on Oscar's face he stopped laughing immediately. "I am sorry, that wasn't at all meant the way it must have sounded. I'm not a doctor, although my 'patients' also desperately need my expertise. My specialty is antique pipe organs—I service and repair them."

Oscar's irritated look softened and now he, too, laughed and raised his glass. "Here's to the different organs in our lives!" he said cheerfully and took another sip.

Only now did Jerry realize how open to misinterpretation and potentially tasteless his remark must have seemed. He wanted to change the subject. "Can I get you another Bloody Mary?"

"No, thank you; anyhow, this is only tomato juice. I couldn't tolerate alcohol at this time of day. I would immediately fall asleep, and my liver would throw a temper tantrum." For some minutes, the two men sat in silence before Oscar picked up the conversation. "How long have you been here?"

"I arrived just over two days ago and went directly to the museum, which has a beautiful but sick old pipe organ. I worked on my 'patient' for almost twenty-four hours. Luckily, some organ parts didn't arrive until yesterday, so I got some sleep while waiting for them. But all in all I can't have slept for more than four hours over the past couple of days—never can in these situations."

Now Oscar was the one to react surprisingly. He violently shook his head while grinning widely. "These coincidences are becoming ridiculous. I also arrived just over two days ago. A rather complicated transplantation was scheduled for yesterday morning, but the donor liver arrived more than twelve hours late. So I was on call and on the phone all that time, and after that I had to perform a long and complicated operation. I hardly got any sleep either."

Jerry looked at the empty glass in his hand and felt the gin rushing to his head. He wanted to put the glass down on the table in front of him but missed its edge by half an inch. The glass fell on the floor but, thanks to the carpet, didn't break. He picked up two slices of lime and the remnant of an ice cube, put them back into the empty glass, and placed it on the table. But while pulling back his hand, he knocked over the bowl of peanuts, which neither of them had touched since Oscar had had his one and only. Jerry's apparent good mood had totally vanished. He now looked deeply depressed. "Maybe she was in such a foul mood because I got her out of bed when I called. I should be at home more, then things might possibly improve between us."

Jerry reached for his mobile and punched in their home number. He never stored the important numbers because he was convinced that his brain would gradually lose its memory capacity if he were to completely rely on speed-dial buttons. But now he had to key in the number several times, and eventually had to look it up. When he finally got through, no one answered. His depression seemed to be getting worse by the minute. "Our lifestyle is definitely not good for the family—that's for sure," said Oscar. He had the feeling Jerry needed to be cheered up a bit. "I always try to look on the bright side. The short sleep I got this morning, for example, was one of the first in weeks that wasn't interrupted by an asthma attack. I had a hefty one this afternoon, though."

Jerry decided that the G&T had been a lousy idea. He still wasn't hungry, but he had to give his stomach something to work on besides alcohol, and he definitely needed some caffeine. "I see that they have croissants over there. I'll get myself a cup of coffee to go with it," he said. "I'll also try to find out whether there are any new developments. What about you? Can I get you something?"

"I don't think I've eaten for twelve hours or more. So, yes, that's an excellent idea. Let's have breakfast and get ready to go home!" exclaimed Oscar.

What does the two men's conversation tell us about the body clock? Where are they, and how can we make sense of the clues they provide? Let's review the evidence—you may want to do so on your own before continuing.

The story begins with a statement about time of day—it is seven o'clock in the evening—and ends with Oscar's exclamation, "Let's have breakfast!" This blatant discrepancy may have put you on the trail. What scenario reconciles this apparent contradiction? You already know that the distribution of chronotypes in a population can be so wide that the extreme larks and the extreme owls are up to twelve hours apart. Could that be an explanation for having breakfast twelve hours out of synch with local time? Probably not; although the most extreme chronotypes are twelve hours apart, they are generally only six hours out of synch with the rest of the world— six hours too early or six hours too late. So the issue of chronotypes is most likely not the basis for the two men's desynchrony with their local time (just for the record, Oscar is a definite early type and Jerry is more of an owl).

Based on what you have read in the previous chapter you might speculate that the two gentlemen are having their conversation on another planet, possibly on Neptune with its eighteen-hour day. A short day like that would make their chronotypes so late that the gentlemen's craving for breakfast at seven in the evening would be conceivable. But then you know that in that story WASPS never sent humans to Neptune. You would also correctly discard the possibility that the two are on Mars because you know that in that situation they would have breakfast at some ungodly early hour. Perhaps the two are subjects who have just finished a long experiment in the Andechs bunker, putting them completely out of touch with local time. Yet both state they have recently finished critical tasks in their profession—one in a museum, the other in a transplant center.

It seems quite apparent that Oscar's and Jerry's internal time is completely out of synch with their external, local time. The simplest explanation, and one familiar to many of us, is that they have just travelled halfway around the globe, so that their biological clocks haven't had a chance to catch up yet. In support of that supposition, they both mention that they arrived only a couple of days ago. But where are they, and to what time zone are their biological clocks still set? From their conversation, we can conclude that their home base is the United States. So the question becomes: what country is about twelve time zones away from the United States?

Oscar and Jerry are sitting in the business-class lounge at Narita Airport in Tokyo and have just learned that their flight to Boston will be delayed by an as-yet-unforeseeable time. People strike up conversations for the oddest reasons, but fellow travelers, stranded and headed to the same destination, have so much in common that it would seem almost impolite not to communicate. They had overheard each other's telephone conversations, and Oscar was not able to avoid hearing Jerry's big row with his wife. The delay would not get Jerry back in time for the dinner party, and his wife didn't hold back her frustration. It wasn't the first time his work schedule had conflicted with their social engagements. She completely ignored the fact that her husband was suffering the worst possible kind of jet lag—a complete half-day out of synch (the reason why Jerry got hold of his wife just before she was about to wake up).

In our modern mobile world, millions of people have suffered from jet lag at least once when visiting friends or taking vacations in distant countries. People whose jobs take them to many different parts of the world find themselves in this rotten state with sad regularity. But what symptoms characterize jet lag a bit more accurately than just feeling lousy? The most conspicuous symptom is tiredness. However, that state is not necessarily specific to jet lag. If someone travelled from Helsinki to Cape Town, he or she might also feel severely tired without ever having left the time zone, merely due to the length of the exhausting trip (not to mention that this traveler would also journey from winter to summer, or from autumn to spring, or vice versa). The major difference between travelling long distances within time zones as opposed to across time zones is that the passenger from Helsinki would have no problem sleeping at the right time of night in Cape Town and would thus recover quite readily from the exhausting voyage. But just travelling long distances across time zones does not necessarily throw the traveler into the state of jet lag.

Long ago, before airplanes carried travelers across the Atlantic, they may have suffered from seasickness but certainly not from jet lag. As the word says, it takes a jet to elicit this state because the main cause for this syndrome is the speed at which we travel from one time zone to another, from one time of dawn to another, from one time of dusk to another, and from one different social timing to another. Our body clocks can cope with the slow changes of dawn and dusk that we experience when travelling by ship. If we were to take a boat from Europe to America, it would be a bit like living on Mars: every evening the sun would set and every morning it would rise a bit later, making the days longer than twenty-four hours. Of course, we would much more easily synchronize to these longer days if we spent some time on deck—especially in the evenings—thereby exposing ourselves to bright light at a time of our body clock when light expands the internal day. If we were to take a boat in the other direction, from America to Europe, we would live on yet another planet with days shorter than twenty-four hours by travelling against the rotation of our globe. If the boat didn't cruise too fast across the Atlantic, these shorter days would still be within the range of entrainment of our body clock, so that our internal time would arrive in synchrony with the external time. According to the principles of entrainment, we would become slightly earlier chronotypes when travelling west and slightly later chronotypes when travelling east.

However, when it takes us less than a day to travel across half of the globe, our body clock is left behind. The shortest flight from Boston to Tokyo takes fifteen hours and twenty-five minutes. If Oscar and Jerry had left Boston's Logan Airport at 8 A.M., it would already be 9 P.M. in Tokyo, and they would arrive at Narita Airport half an hour after noon local time (it would actually be a whole date-day later since they would have crossed the date line). Their internal time would, however, be set to thirty minutes before midnight—approximately half a day out of synch.[1]

Another symptom of jet lag is nighttime insomnia despite utter exhaustion. Based on the calculations above, this is not surprising. Unless we have travelled to a holiday destination, we are expected to be active when our body clock is on its way to bed, and we have to try and catch up on sleep when our internal alarm clock "announces" that it is time to get up. As a rule of thumb, it takes the body clock approximately one day per travelled time zone to adjust to the new cycle of light and darkness, so that Bostonians travelling to Japan would need approximately twelve days until they functioned normally again. The difficulty of not being able to sleep adds to the exhaustion of the trip itself. Since this state may continue for many days after arrival at the destination, the traveler cannot compensate for the exhaustion experienced during the long flight.

Holidaymakers are often advised to live as much as possible according to the normal day at their destination if they want to get rid of their jet lag as quickly as possible. This includes being active and seeking daylight when the sun is up, and resting in darkness at the usual times of the host country, even if normal sleep is impossible during the first few days after arrival. It also means avoiding taking too long a siesta in the afternoon. Siestas roughly correspond to bedtime at home, making it even less possible to sleep through the new local night. The stronger you can make the new zeitgeber, the faster your body clock will arrive at the new destination. The rule of thumb about our body clock needing one day for each shifted time-zone hour is an approximation. Most people can adjust more easily when flying to the west. Going west, their clock adapts by expanding their internal days, which are already longer than twenty-four hours. Accordingly, adjustments after travelling in the opposite direction are more difficult, especially for late chronotypes. Only extreme early types report adjusting faster and more easily when travelling from west to east.

Other symptoms of jet lag are reduced alertness, poor motor coordination and reduced cognitive skills.[2] You might think that these symptoms go with the territory when we suffer from exhaustion and profound sleep deprivation. But again the combination of being active at the wrong internal time and being exhausted and sleep-deprived exaggerates our impairment. Cognitive states (like vigilance, alertness, and attention) and skills (like motor coordination, performing simple calculations, or memory tasks) are as much under the control of the body clock as sleep and wakefulness, body temperature, or circulating hormones.[3] The Constant Routine experiments, during which subjects are tested around the clock without sleeping or moving around much, have shown that subjects perform worst at many skills sometime during the second half of the internal night and improve thereafter, despite having been awake for an even longer time.[4]

Not being able to properly catch up on our sleep and our exhaustion is not the only problem we face during jet lag—sudden mood changes and even depression are sometimes also symptoms.[5] Another important issue during jet lag concerns our appetite and our digestion. Both are controlled by the body clock, so that during jet lag we get hungry at times when we should try to find sleep, and we have no desire to eat when the locals normally do. If we force ourselves to eat, we confront our stomach with food when it cannot produce enough of the juices necessary for normal and efficient digestion. At the other end of the day, we might lie in bed trying to sleep but feeling hungry. Our stomach anticipates food and produces digestive juices that can then act only on its empty self—an ideal condition to support the development of ulcers.

Our brain and our digestive tract seem to suffer most. Or perhaps they are those parts of our body that give us the most obvious problems during jet lag. Scientists at the University of Virginia submitted rats to conditions that simulated jet lag and found that the SCN, the master clock in the brain, apparently adjusts much faster to a new light–dark cycle than peripheral tissues like muscles, lung, or liver, shown in the rhythm of wheel-running activity.[6] In a similar series of experiments, the Virginia researchers investigated what happens if they put animals on restricted feeding programs.[7] In the control conditions, they gave animals access to food during the night, corresponding to the normal activity time of the nocturnal rat. After a week, they shifted the feeding times by several hours without changing the times of the light–dark cycle. The results of these experiments suggest that the clock in the liver can be entrained by food, while the master clock in the brain, the SCN, continues to entrain exclusively to the light–dark cycle. Under these conditions, the normal circadian harmony of the different body clocks is pulled apart.

The temporal disharmony of different body parts may be an important cause of our lousy feeling in the state of jet lag. The Virginia experiments also indicate that it may take different organs a longer or shorter time to adjust to the new time regime at the destination.

Jet lag sufferers can therefore justifiably ask the question, "When will my organs arrive?" Most chronotypes adjust more easily after a westward flight, although some rare early types adjust better in the other direction. Imagine that the different organs of our body showed similar differences in adjustment, some by expanding and others by compressing their internal days. In that case, it would even be conceivable that while Oscar and Jerry flew from Boston to Tokyo on a westward flight via California, some of their organs virtually had to fly the other way around the globe, via Paris and Moscow. Until all the different organs "arrived" at their final destination, it would not be surprising to feel lousy.

16

After graduating from high school, Timothy had visited Benjamin at Princeton University and had worked in his friend's small start-up company. Ever since they lived on the same street in a suburb of Eugene, Oregon, they had been the closest friends, even though Benjamin had been the classmate of Timothy's elder brother. The six months in Princeton had been Timothy's first taste of independence away from home and had been heaven on earth. He loved Princeton, with its university and its sidewalk cafés.

Benjamin had opened a small shop called LayIn&Out—a sophisticated kind of copy shop and café. Besides providing all the normal photocopying possibilities, customers (predominantly students) could bring documents or slide presentations for professional improvement on layout. Benjamin had gotten the idea after watching an Indian Bollywood movie that told the story of an old-fashioned "letter writer" who provided communication services in a faraway little village where only a few people could read and write. Afterward in the bar, he discussed with friends how this profession still persisted in other ways in modern society, for example when experts advised customers on layouts for their documents, presentations, or personal websites.

Besides being an excellent coffee bar, LayIn&Out offered several desktop computers, each of which had a Skype connection to one of Benjamin's young employees, who had to be on call when the shop opened at seven in the morning. The early morning hours had the highest customer traffic because many students worked all night on a paper or a presentation due the next day. To give it the final polish, they took it to LayIn&Out for professional improvement while trying to compensate for the sleepless work night with an excellent latte and a delicious croissant.

Timothy was good at polishing layouts and had always been an early type compared with the rest of his peer group. So, during the six months of his Princeton visit, he quickly became an irreplaceable cog in the up-and-running machinery of the young company. Although he loved his life in Princeton, he decided to return home after half a year. Amy and Timothy had always been very close, but it took the long separation and endless Skype sessions for them to realize that something more than friendship had developed between them. Benjamin insistently but unsuccessfully tried to persuade Timothy to stay—his role at LayIn&Out was crucial. In the end they worked out a compromise that involved frequent visits to Princeton (billed to the company's expense account) and a continuation of Timothy's expertise—online from Oregon. So Timothy went back home with a well-paid job and a brand new super-fast computer in his luggage. From the day of his arrival in Eugene, Timothy sat at his computer every work day (LayIn&Out was closed on Saturdays), advised customers via Skype, worked on their documents' layouts, and delivered the final product in the shortest possible time.

Another half a year into the existence of LayIn&Out, it became clear that the concept was a terrific success and that the young company was making a huge profit. The only thing standing between Timothy and complete happiness (including Amy, of course) was his working hours. He got up at 3 A.M. every work day to be ready, and sort of awake, in front of his computer at four—just in time for the shop's opening time in Princeton. He then worked hard and concentrated for at least six hours without interruption. The consequence of this strict and early schedule was chronic exhaustion—he never got enough sleep, despite his early chronotype. Timothy could catch up on his sleep only on weekends by sleeping in a completely dark room. But he never slept beyond eight, even though that was much later than he would usually wake had he not been so exhausted. His relationship with Amy began to suffer seriously because he tried to be in his dark bedroom at around nine every evening, although he never could get to sleep until about 10 P.M. His best breaks from this gruesome regime came on his frequent visits to Princeton, where he would participate in meetings that tried to improve LayIn&Out's policies and concepts. Even then, he worked as one of the on-call partners who were available for the customers via Skype at seven in the morning. But at least he was getting enough sleep and felt much fresher during the day compared with his online participation in the scheme from Oregon. This difficult situation went on for about another nine months until the company was so successful that they decided to open branches across the country. Eventually, Timothy became the West Coast manager of LayIn&Out with his own stores in his own time zone. He still continued to work as an online representative, but the relaxed working hours gave him so much more energy that he had no trouble combining his double role as online layout expert and West Coast manager while still finally finding enough time to spend with Amy.

At the beginning of this book, I covered a lot of the biology underlying the ticking of the body clock. But as the book progressed you have read more and more about the interferences between body clock and social clock—between internal and external time. The last chapter focused on jet lag and explained what happens when we travel too fast across time zones. This chapter looks more closely at the discrepancies between internal and external time that can occur even if we never board an airplane and stay in the same place for our entire life.

You meanwhile know that sleep duration and sleep timing (chronotype) are two independent, heritable traits. However, this is only true if we consider the average sleep-need of an individual, calculated across both work and free days. The independence of the two traits breaks down if we analyze work days and free days separately: sleep duration can be very different, and this difference very much depends on chronotype. The later someone's chronotype (horizontal axis of the graph; the vertical axis represents sleep duration), the less sleep this individual gets during the working week (black dots). There is a simple explanation for this observation: later chronotypes fall asleep later than early chronotypes—even on work days. Yet working hours are, on average, the same for everyone, which is why most people use an alarm clock to wake up in time for work. (Our database shows that this is true for 85 percent of the population.) Consequently, the later someone's chronotype, the shorter his or her opportunity for sleep on a work day.

Sleep duration on free days (open circles in the graph) reflects the systematic sleep deprivation that different chronotypes accumu-

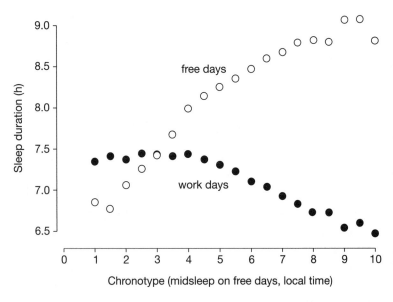

The scissors of sleep. Depending on chronotype, sleep duration can be very different between work days and free days.

late on work days. Very late chronotypes don't get much more than 6.5 hours during workweek nights and therefore sleep for nine hours or more on their free days. Some individuals have to sleep through half of their free days to catch up on their sleep loss.

In contrast to later chronotypes, early larks experience sleep deprivation on free days. Again the explanation is simple: the majority of later chronotypes in the population puts pressure on early types. Evenings preceding free days are especially difficult: don't be so bloody boring; come with us to the bar, the cinema, the theater . . . Late and early chronotypes almost lead mirror lives between internal and external timing. The body clock tells late types when to fall asleep, and the alarm clock tells them when to wake up. In contrast, social pressure exerted by their owlish friends tells larks when they are allowed to fall asleep on evenings before free days, and the body clock tells them when to wake up. It almost doesn't matter when early types go to bed—they more or less wake up at their usual time the next morning.

To validate how accurate and representative the answers of subjects are when they fill out the Munich ChronoType Questionnaire, we periodically ask people to keep sleep logs for at least six weeks. Our current collection of these logs totals more than a thousand. We find that the correspondence between what people state in the questionnaire and what they state in their daily sleep logs is almost perfect. These logs not only allow us to validate the questionnaire, they also tell us very personal stories about the temporal life of different people. When we sort these sleep logs according to chronotype, clear patterns of sleep behavior across the working week and the weekend become apparent. Let's look at three representative examples. In all these examples, sleep on work days is shown as black bars, and gray bars represent sleep on free days. The successive days of the self-recorded sleep times are shown from top to bottom (vertical axis), and local time—from 6 P.M. to 6 P.M.—is indicated on the horizontal axis.

The first example represents the sleep times of an extreme late chronotype who can freely choose her own work times. This individual falls asleep, on average, between 3 A.M. and 4 A.M. and wakes up between 11:00 and noon. Neither sleep timing nor sleep duration differ significantly between work days and free days; both sleep onset and sleep end show a natural day-to-day variation. As you will see,

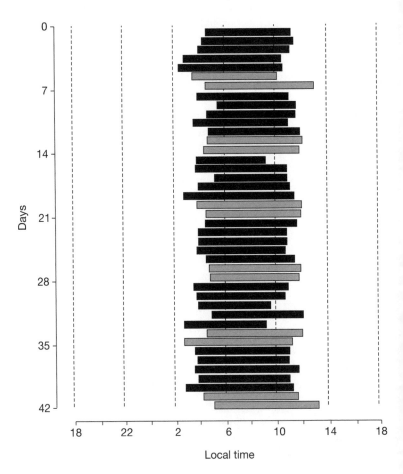

The sleep times of an extreme late chronotype who can freely choose his own work times. Black bars represent sleep on work days, and gray bars represent sleep on free days.

this is the least constrained sleeping situation of all. This individual sleeps within her body clock's window for sleep practically all the time.

The subject in the next example is an extreme early type who goes to bed sometime between 8 P.M. and 10 P.M. and wakes up between four and five in the morning—long before workplaces usually

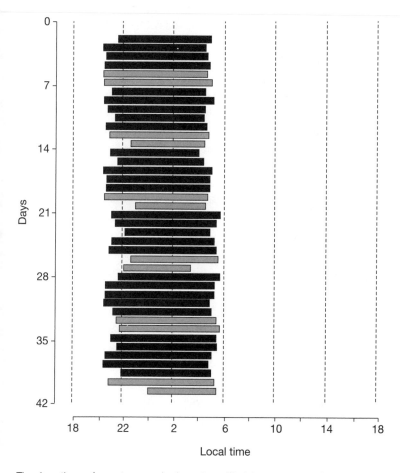

The sleep times of an extreme early chronotype. Black bars represent sleep on work days, and gray bars represent sleep on free days.

start the day. I am always amazed about the huge differences that individuals show in their sleep–wake behavior. The individuals who produced the sleep logs in these first two examples could almost share a bed without ever having to sleep together. As in our first example, sleep onsets and sleep ends of the early type show a natural day-to-day variation. In this case, however, a systematic pattern on weekends appears: early types become sleep deprived on free days as a consequence of the social pressure exerted by their owlish friends, who are the majority.

Typical for a bell-shaped distribution, extremes at both ends are rare. Extreme late types who can freely choose their work times may be even scarcer—work times are too early for 60 percent of the population. Thus, the majority has difficulty complying with given working hours. The "scissors of sleep" show that the later the chronotype, the greater the differences in sleep duration between work days and free days. But sleep timing can also show huge differences between the working week and the weekend in late chronotypes. The scalloped shape in this third sleep-log graph is far more typical for the majority of the population than the two first examples.

Sleep onsets on work days show the usual day-to-day scatter. Sleep ends on work days, however, fall—with some exceptions—onto a straight vertical line that can only be explained by assuming the regular interference by some kind of waking-up aid (WUA; I tried to invent some positive-sounding term for the obnoxious little gadget called the alarm clock).[1] The use of a WUA in the morning depends, of course, on chronotype. While few larks use a WUA on work days, almost all owls do. Modern WUAs, which are equipped with a "snooze" function, would give us an even more quantitative characterization of just how chronotype-dependent waking-up rituals are. I am convinced that if we were to query people on how often they use the snooze button on workday mornings, we would get a perfect curve, with extreme early types never making use of this possibility and extreme late types using it many times before finally getting out of bed. The fact that many early types actually use an alarm clock

(despite always waking up before it rings) must lie in some un-
founded paranoia about not waking up in time for work.

The most characteristic feature of the sleep–wake behavior of
late chronotypes is the strange scalloped shape of their sleep times.
Their large shifts are somewhat reminiscent of what Oscar and Jerry

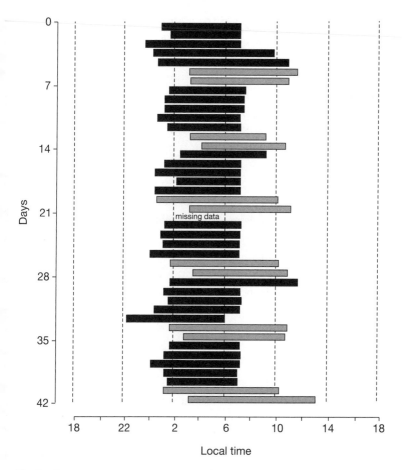

The "scalloped" sleep–wake behavior of late chronotypes who use an alarm clock to
wake. Black bars represent sleep on work days, and gray bars represent sleep on
free days.

experienced by flying from one time zone to the other. The patterns in the sleep logs of later chronotypes almost look as if these individuals had flown from the East Coast to the West Coast of the United States on Friday, and then back again on Monday.

I am often asked whether we cannot get used to given working hours merely through discipline and by confining our sleep habits to certain times. The assumption inherent in this question is that the human body clock can synchronize to social cues. I tend to find that any such questioner, who usually also displays a somewhat disdainful tone toward the weakness of late chronotypes, is an early type—someone who has never experienced the problems associated with the scallop-shaped sleep–wake behavior of late chronotypes.

The story of Harriet and her golden retriever made it quite clear that social cues can be ruled out as a good zeitgeber. Otherwise blind people would certainly use these cues to relieve themselves of the misery of having a body clock that runs at its own pace through a twenty-four-hour society. Social cues have been shown to entrain human clocks, but notably only in sighted people. Oil rig workers are a good example. They work offshore for several weeks at a time, during which they work for most of their active day before retreating to their bunks. They can be put on almost any shift schedule without problems because their body clock readily entrains (as long as the work schedules don't rotate too fast). Yet their body clock is only entrained indirectly to the social times of their shift. The main zeitgeber is still the light–dark cycle, which they create themselves by being exposed to light during work and to darkness while they sleep. Note that despite the synchronization of the body clock to socially dictated light–dark cycles, the phenomenon of early and late chronotypes remains—only now in relationship to the social schedule rather than in relation to the day-night cycle on our planet. The phase of an individual's body clock in relationship to a zeitgeber is a biological phenomenon and not a matter of discipline.

The circadian clock in normal workers is rarely oblivious to daylight. That is why their body clock entrains to the natural alterations of day and night and not to social work times. Many of the sleep logs in our collection that display the typical scallop pattern were kept by individuals who have been working with the same early and regular work hours for decades without ever getting used to them. As a result, these individuals appear to fly west across several time zones every weekend, returning east every Monday. Because this pattern is so much like time-zone travel, we have introduced the term *social jet lag.* Unlike what happens in real jet lag, people who suffer from social jet lag never leave their home base and can therefore never adjust to a new light–dark environment, as Jerry and Oscar could have done if they had only stayed long enough in Japan. While real jet lag is acute and transient, social jet lag is chronic. The amount of social jet lag that an individual is exposed to can be quantified as the difference between midsleep on free days and midsleep on workdays (the change in midsleep in the graph below, which is based on the third sleep-log example). Over 40 percent of the Central European population suffers from a social jet lag of two hours or more, and the internal time of over 15 percent is three hours or more out of synch with external time. There is no reason to assume that this would be different in other industrialized regions.

The condition of social jet lag is comparable to having to work for a company that lies several time zones to the east of one's residence. Timothy the layout expert is an early type. He could therefore sit fresh as a daisy in front of his computer when LayIn&Out opens its doors at seven o'clock. But that was true only when he lived in Princeton, New Jersey. Once he had to supply the same services out of Eugene, Oregon, he had serious difficulties in being awake and ready for work. Timothy's story shows even early types the difficulties that a large portion of our society has to face every morning when trying to get ready for work. The main symptom of social jet lag is chronic sleep deprivation, which can be the cause of many health and mood problems. We are only beginning to understand the potentially detrimental consequences of social jet lag. One of these has already been worked out with frightening certainty: the more

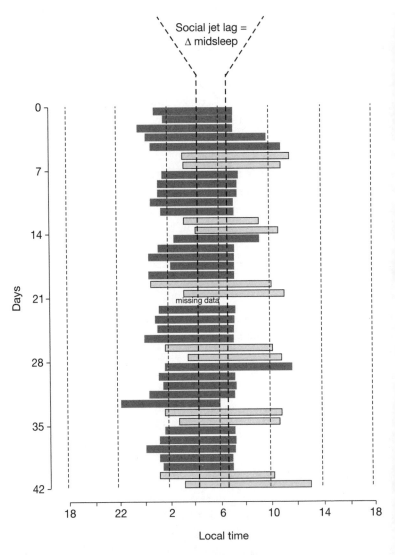

Social jet lag is defined as the difference between midsleep on work days and midsleep on free days. This definition allows quantification of the discrepancy between an individual's social and biological timing. Dark gray bars represent sleep on work days, and light gray bars represent sleep on free days.

severe the social jet lag that people suffer, the more likely it is that they are smokers. This is not a question of quantity (number of cigarettes per day) but simply whether they are smokers or not. In relation to smoking there are only three types: those who never smoked, those who have successfully quit smoking, and those who still smoke. Statistically, we experience the worst social jet lag as teenagers, when our body clocks are drastically delayed for biological reasons, but we still have to get up at the same traditional times for school. This coincides with the age when most individuals start smoking. Assuredly there are many different reasons people start smoking at that age, but social jet lag certainly contributes to this risk. Later on in life, the less stress smokers have, the easier it is for them to quit. Social jet lag is stress, so the chances of successfully quitting smoking are higher when the mismatch of internal and external time is smaller. The numbers connecting smoking with social jet lag are striking: among those who suffer less than an hour of social jet lag per day, we find 15 to 20 percent are smokers. This percentage systematically rises to over 60 percent when internal and external time are more than five hours out of synch.

17

Olaf Toemmelt is the director of GoEast, an extremely successful public relations agency based in Magdeburg, East Germany. The State of Saxony-Anhalt has launched a competition for a public relations campaign that would help to attract more businesses and boost the state's economy. Together with his team, Olaf had been brainstorming for almost a week now, but none of the ideas had been convincing enough to develop into anything that could be used for the competition. It was a beautiful summer morning, and he drove to work in his open BMW convertible from his country house—an old vicarage he had acquired for an extremely reasonable price just after Germany's unification—along the highway toward Magdeburg. His blond hair—he had seen his hairdresser only yesterday—was protected by an original NY Yankees baseball cap from the 1930s that he had bought at an auction on one of his frequent visits to the United States. His thoughts were occupied with the still-unborn PR campaign, so he was only half-listening to the morning news program. "A new poll was published yesterday," the radio announced, "investigating when Germans get up. It showed notable differences among the different German states." Olaf was still not listening properly. "On average Germans get up at 6:48; the latest out of bed are the citizens of Hamburg, Berlin, and the state of Hessen; among the earliest are Thüringen and Saxony. The all-German champion of early risers is Saxony-Anhalt, where people get up nine minutes earlier than the German average." "Ha," thought Olaf, "they get up earlier than the rest of the country—so what? That's only because the 'Ossies' still live by the rules of the German Democratic Republic. They have had to get up early all their lives—it was party line."[1]

The traffic in front of him came to a standstill. "Darn it," he exclaimed, and continued less loudly, "I'll be late for the morning meeting, and I will get in after Brigitte, that lazy woman. Just imagine the look on her face if I walk in after her." The traffic started to move again, on a stop-and-go basis. "Getting up early also means retiring early. You can see this in town, where everything closes before people in Berlin even get ready to go out for the evening," he thought. "Berliners don't get up early and still attract good business." At that point, the traffic jam dissolved as quickly as it had formed. About fifteen minutes later he parked his car in front of the building where GoEast had rented a suite of offices.

When he entered the conference room everyone was already there, including Brigitte, who gave him a broad smile. Olaf opened the meeting. "Has anyone had a breakthrough idea for the state PR campaign overnight?" To his chagrin Brigitte was the first to raise her hand.

"Did anyone hear the cool story on this morning's news?" she asked. "It was about a study which found that the people of Saxony-Anhalt are the first to be up every morning."

Olaf waved his hand in a dismissive gesture: "Yes, I heard it. It's quite pathetic, really."

"But on the contrary," replied Brigitte, "Olaf, leave behind your 'Wessie' prejudice and think of how you would have reacted if they had found our state to be the last up every morning!" She had always thoroughly disliked her boss's two-faced attitude. On the one hand he thought that East Germans were a bunch of losers, and on the other he was one of the first to realize that there was a lot of money to be made in the East. Not to mention the fact that he had bought the most beautiful old mansion for a bag of peanuts. "Your reaction would have been similarly negative. 'Typical,' you would have said, 'who would have thought anything else?'" Brigitte now had the full attention of everyone in the room.

"This is our big chance! This morning's newscast has solved all our problems. It has delivered our campaign slogan on a silver platter. Here is what I suggest: 'Saxony-Anhalt—we get up earlier!'" In an instant, a whole firework of ideas, images, and action items exploded in the morning meeting. Getting up early is more than just getting out of bed—it is a mentality that is reflected in the state's long and successful history. Getting up early is the mindset of a state that is eager to be the best, ready to move ahead. Saxony-Anhalt is Germany's early bird that catches the business.

Even Olaf started to see the posters and the TV ads. This was it! There was no doubt that his company would win the competition, and that his idea, his campaign would change the face of Saxony-Anhalt. Olaf was convinced that he had brought the campaign's central idea into this morning's meeting, and he was very proud. Still, from then on, he was oddly hesitant to treat Brigitte with the same disdain that he had ever since this Magdeburg-born woman had started to work for GoEast.

When you drive on the highway from anywhere in the southwest of Germany toward Berlin, you quite likely will pass through the state of Saxony-Anhalt. The signs announcing the entry into this state carry the slogan, "Saxony-Anhalt—welcome to the country of early risers."[2] This campaign has been the cause of many comments and jokes on the internet. One of them says, "If I still lived in Saxony-Anhalt, the early-riser record would collapse since my daily habits would completely mess up the average." Another one refers to the fact that workers in Saxony-Anhalt had, on average, such a long commute to their workplace compared with workers in other German states that they simply had to get up earlier. A third comment, written by someone who had recently moved there, asks: what makes the people from Saxony-Anhalt rise earlier? And someone answered with a smiley: "They think of their dogs, who want to go and 'read' their 'morning paper' at the crack of dawn."[3] This joke probably comes closest to the truth.

The poll mentioned in our story is real. It had apparently asked two thousand Germans when they get up. Although this poll is not very representative (it averages to no more than 125 individuals per federal state), its results are remarkably close to what you will read about in this chapter. A couple of years ago, I approached the question of whether the human clock is synchronized by the light–dark cycle—like the circadian clock of other animals—or whether humans are an exception whose clocks are predominantly synchronized by social cues. In a first approach, I conducted an experiment together with colleagues in India. We collected thousands of chronotype questionnaires in two cities: Chennai, on the east coast of the Indian subcontinent;[4] and Mangalore, a city at the same latitude on India's west coast. The first results showed that the average chronotype in Chennai is clearly earlier than that in Mangalore, indicating that the light–dark cycle is the main zeitgeber for the human body clock—the sun rises earlier in the east than in the west. When I presented these preliminary results at a conference in Florida, a British colleague came up to me after the session and said, "Beautiful data, but I bet you this is *India*—I'm sure you won't find anything like this in Europe!"

I was quite shocked by this remark, but I owe my colleague a debt of gratitude for making me look for a way of proving him wrong. At home I had a database of chronotypes in Central Europe that contained at that time approximately 40,000 individuals. Since we had asked the participants to enter the postal codes of their current residence, it would be possible to create a map of Germany based on the coordinates corresponding to the postal codes. Back home, I immediately began to analyze the database and generated the map. When I finally looked at the finished product, I was astonished to find how well the database covered the entire country.

I decided to use only German entries for the analysis and not other countries for two reasons.[5] First, they were by far the biggest contingent, and second, I would not have to defend the results against

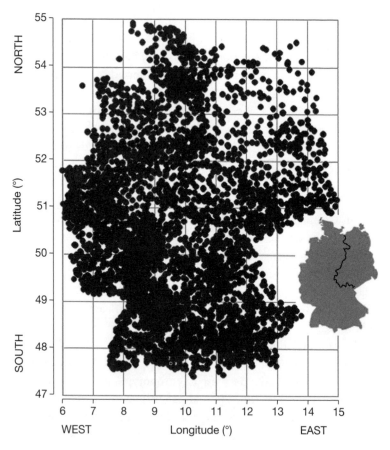

Entries in the Munich ChronoType Questionnaire database represent all of Germany well. Reprinted from T. Roenneberg et al. (2007). The human circadian clock entrains to sun time. *Current Biology* 17(2):R44–R45, with permission from Elsevier.

a critic who pointed to potential cultural differences. Germany is a big country and certainly not very homogeneous in its culture, but Berliners and Bavarians resemble each other more than, say, people from Baden-Baden and from Strasbourg. Each of the points shown in the map represents up to several hundred entries. Their underlying number strongly correlates with the population density of the respective location. There is no need to draw borders around the area of postal code coordinates—Germany's shape is clearly represented (see small inserted map in gray). There are even some entries from people living on German islands in the North Sea.

Based on this result, all I had to do was cut Germany into longitudinal slices from east to west and determine the average chronotype in each of them. Each longitude represents one of the 360 degrees of the earth's circumference, and the earth makes a complete turn every 1,440 minutes. The sun therefore takes four minutes to pass over each longitude. At its widest point, Germany extends over nine degrees of longitude, so that the sun rises thirty-six minutes earlier at the country's most easterly point compared with its most westerly one. If the body clock of humans were entirely influenced by social time, all Germans should have, on average, the same chronotype at all longitudes because, unlike in bigger countries like the United States, Canada, or Russia, Germans all live in the same time zone. If, however, the human body clock were entrained by the light–dark cycle—by dawn, dusk, or simply by the sun's highest inclination (at solar midday)—then the difference in chronotype should be four minutes later in every longitude from east to west. Because of all I knew about circadian entrainment in all kinds of different organisms, I would have bet that sun time also has an influence on the timing of humans. However, the clarity of our results was surprising, even to me. In the graph below, the horizontal axis represents the local time of sunrise for each longitude on the longest day of the year (this is just a point of reference—I could have chosen any other day of the year). The vertical axis shows the local time of midsleep on free days (representing chronotype) averaged for all people in our database who live in a given longitudinal slice of Germany. The dashed diagonal line represents the east–west progress of sunrise. You can easily see that chronotype moves with the sun, becoming later by, on average, four minutes per longitude from east to west. Note that sun times are inverted on the horizontal axis, increasing from right to left, so that west is to the right and east is to the left, as they are on maps. These results show that the body clock of humans is no different to that of animals. It entrains to sun time—even in Europe.

It is a remarkable fact that every time I find evidence for a biological basis for human behavior, I am confronted by the conviction that the same results could be explained far better by cultural causes.[6] The first time I presented these results, I was instantaneously confronted with the claim that this is a cultural artifact! The people in the former GDR always had to get up earlier than their fellow coun-

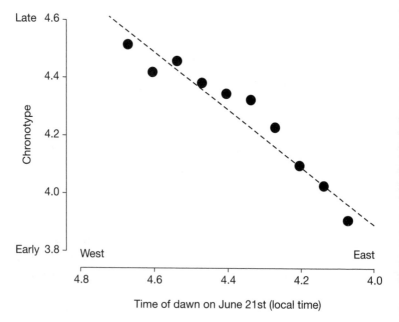

Chronotype in Germany moves with the sun, becoming later from east to west.

trymen in the West.[7] The social-explanation camp claims that our study merely shows a remnant of the cultural differences between people having grown up under either a socialist or a capitalist regime.

If this were true, the results should have looked more like what is shown in the figure below. Depending on the "mixture" of capitalists and socialists living in each of the longitudinal slices, the averaged chronotype should fall onto an S-shaped curve, with pure East Germany on the right, pure West Germany on the left, and some mixture of the two in the middle.

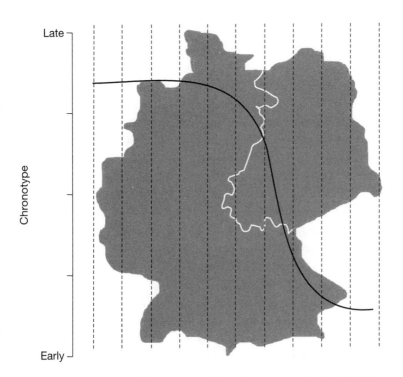

The German data show that the human clock is predominantly synchronized by sun time rather than social time. The curve here predicts average chronotype from west to east if social time was the dominant zeitgeber. In reality, average chronotype for each degree of longitude falls on a straight line, as shown in the previous figure.

However, the real data form a straight line parallel to sun time, which weakens the purely sociopolitical hypothesis. But to be sure that the systematic east–west changes in chronotype are not a socio-political artifact based on the transition between the former GDR and West Germany, we simply analyzed the southern regions of Germany, which have almost the same east–west expanse but were never ruled by a socialist government.[8] The results were practically identical to those for the entire country. So, we finally were able to convince our critics that the human clock is predominantly synchronized by sun time rather than social time.

As you read in the introduction, time-zone time is the temporal reference that people have lived by since the late nineteenth century, when the world was subdivided into twenty-four time zones. Before that, the temporal reference was local sun time. It is quite remarkable that we find—in the first part of the twenty-first century—that our body clocks still live very much like those of our ancestors, namely by sun time, while our entire social life has to conform to a different schedule. We might think that—since each time zone is only one hour apart from the next—we would have to live, at most, only one hour away from the actual local sun time. And if a time zone spreads equally to the west and to the east, then the forced deviation would be only thirty minutes, which seems like an acceptable difference. After all, we change our reference time by a full hour twice a year when we change to and from daylight saving time.

Unfortunately time-zone time and sun time rarely differ by just thirty minutes or less because politicians decide to which time zone their country's citizens belong. Before I started to investigate the discrepancies between social and sun time, I was completely oblivious to how different midnight could be from its original meaning of "mid-dark." The most extreme example is China. Its entire mainland territory, which extends over almost a sixth of the earth's circumference, is fused into one single time zone referenced to Beijing sun time.[9] When people in western China look at their watch and see that it is 10:00 P.M., it is actually only 7:24 P.M. by sun time, and if they had to get up at six in the morning to go to work, it would by local sun time be only 3:24 A.M. I have been told that the western Chinese population actually doesn't orchestrate social life by Beijing time. For example, when they come together for an early evening meal (say at around 7 P.M. local sun time), they would arrange to meet at 11 P.M.

Similar though not quite so drastic differences exist even in Central Europe. By definition, mid-dark and midnight coincide, for example, in London, in Beijing, or in Prague.[10] The Central European time zone is ahead by one hour compared with Greenwich, which corresponds roughly to the sun time in Prague, lying 15 degrees east of Greenwich. Thus, midnight occurs almost one hour before mid-dark in Paris and more than ninety-seven minutes earlier in Santiago de Compostela, the capital of the Spanish province of Galicia and the most western city of Spain. The difference between social time and sun time becomes even greater for more than seven months of the year when we live under daylight saving time. Then the time difference between midnight and mid-dark in Santiago de Compostela is as much as 158 minutes—two hours and thirty-eight minutes! When the clock in this Spanish town shows midnight, it is actually only 9:22 P.M. by sun time.

The large difference between social time and sun time in Galicia is especially interesting for our research because this Spanish province is very close to northern Portugal (and at the same longitude), yet the Portuguese population lives by Greenwich Mean Time. For that reason, I am trying to initiate studies that compare daily behavior between Galicians and northern Portuguese. When I mentioned this plan to a scientist from the northern Portuguese town of Porto, he responded without hesitation that such a study would be quite fruitless because of the strong cultural differences between the two populations. To underline his argument, he said, "In northern Spain, for example, they have dinner a whole hour later than we do in Portugal."[11] I am gradually beginning to get some fun out of this sociocentric view. In the case of adolescents, I argue that it is not the disco visits that make teenagers late chronotypes, but that discos are the only sanctuaries in our society where they can be loud at that time of night without waking up the rest of the population. With similarly reversed reasoning one could argue that some cultural differences result from the biology of the body clock. Although all live according to their social time, East Germans have their dinner first, followed by the West Germans, followed by the French, and finally followed by the Galicians. In the next chapter you will see that there are even body-clock reasons why people in the country have their dinner before the town folks.[12]

18

Sophie sat at her favorite spot in the alcove window overlooking the valley. After everyone else had gone to bed, she had switched off the lights to see the beautiful starry night outside. A full moon was just about to climb over the ridge of mountains, and its solar reflections bathed the valley in the most magical light. Her thoughts went back to those terribly confusing months when none of them had any idea about how to solve the gridlocked situation. Before everything had started to turn upside down and inside out, they had been a happy and normal extended family.

Joseph and Frederic were identical twins born thirty years ago on the family farm in the midst of the mountains. After finishing school, Joseph started to work full time on the farm with his parents. Six years ago they decided that it was time for the old folks to move into the small annex house. The farm was tucked away at the very end of a long, dead-end valley, about half an hour by car to the next small town, where Frederic worked in a small factory. The two were typical twins, walking the fine line between wanting to be separate individuals and wanting to do everything together. They had even met their future wives on the same evening at the annual dance of the local fire brigade!

The twins had danced with Sophie and Hanna all night, constantly exchanging partners. The two women were similar in looks and demeanor and shared the same sense of humor. As one might expect, the twins had very much the same taste in women, so it took them months to sort out who loved whom more. In the end, those pairs that seemed to have more in common ended up together. So-phie had frequently gone to the farm to give Joseph a hand, whereas Frederic spent most of his lunch breaks visiting Hanna at the local grocery her parents owned. He then usually gave her a shopping list for the farm's groceries. She had everything packed up by the time he came around in the evening to collect the supplies before driving up the long and winding road to the head of the valley.

A year later, Joseph married Sophie, whose parents owned a farm just twenty miles as the crow flies. Soon thereafter Frederic and Hanna also married and moved into the attic of the farm's main house, which the twins had started to remodel soon after the first wedding. Hanna had always been more of a farm girl at heart, feeling slightly out of place in her parents' grocery trade. They all could have led a normal, uncomplicated, and happy life had it not been for the evenings and the mornings.

Until Frederic started to work in the factory, the twins lived as if controlled by the same clock. At the same times—almost down to the minute—they would start to yawn, to fall asleep, and to wake up (without alarm clocks). But the longer Frederic worked in the factory, the more the twins' timing drifted apart. While the farmer kept to his old habits, the factory worker went to bed later and later and now definitely needed an alarm clock. Much as the young wives had in common, they were almost opposite in their daily habits. Hanna had to be in bed at around eight. She fell asleep immediately and was up well before the sun (except for a couple of weeks during the summer). But Sophie had always been a late type. In his speech at their wedding, her father had praised all her virtues but had warned Joseph how difficult it was to get her out of bed. So it came about that Joseph and Hanna took to tending to the cows in the morning and then making breakfast for the others. At the other end of the day, Frederic and Sophie would be the last to go to bed, long after their respective spouses.

Common life stories and practical issues had helped in making the matches, but now the daily routine started to reshuffle the quartet's emotions. Frederic and Sophie grew closer and closer during their shared evenings. Hanna and Joseph strengthened their bonds during the early morning hours. The changes occurred so gradually that none of them noticed. But one night Hanna woke up, went to the kitchen to make herself a warm cup of milk, and found her husband in the arms of her (and his) sister-in-law. Hurt feelings are the poison of harmony. The incident threw the quartet into a maelstrom of accusations and distrust, made worse by the fact that—over the next months—no one got enough sleep anymore. Hanna and Joseph stayed up well beyond their usual bedtimes so as not to leave their spouses unguarded in the evenings, and Sophie and Frederic made an effort to get up early because they too distrusted their partners.

Hanna fell into a deep depression and stopped commuting into town with her husband to work in the family's shop. At moments when she could think more clearly, she had to acknowledge that she had also started her early mornings joyfully, anticipating her time with Joseph. When his hand accidentally touched hers during work (which had happened more and more frequently), she drew hers back quickly, because the feelings these incidents triggered filled her with guilt. In the end, paradoxically, it was Hanna's worsening depression that had saved them. She hardly left the house anymore and was starting to go to bed as early as six, getting up at two. The other three sat together one evening and Frederic told them about the difficulties Hanna's aging parents were having in keeping the shop running without their daughter's help. "If only I had more time," Sophie said, "I would go and give them a hand." Frederic suggested that Hanna could take over some of her responsibilities on the farm. "She would have to be outside much more, and that would surely also help in fighting her depression," he said. The plan was put into action, and the new working constellations somehow seemed to relieve the emotional tension as well. In the end, the quartet found a solution that was ideal for everyone. From then on, they were a happy extended family once more, one that became even happier when the two couples became—over the years—parents of a considerable number of children.

⧗

Although chronotype can influence a relationship, Hanna's difficulties and Frederic's new sleep schedule illustrate the importance of zeitgeber strength, highly relevant in our modern society. Joseph and Frederic are identical twins who grew up under identical conditions on their parents' farm. So the genetic basis on which their body clocks are built is identical, and their daily behavior was originally completely synchronized.[1] Their temporal habits drifted apart only when Frederic started to work in the factory—he appeared to become a later chronotype. This phenomenon can be explained by the principles of entrainment. Recall the hide-and-seek game of entrainment. To make the length of the internal day the same as that of the external day, the clock has to expose the appropriate portions of its response characteristics in the light and hide the other portions in the dark. I have provided an example of this hide-and-seek game with three theoretical body clocks—the twins', Hanna's, and Sophie's —whose internal days are either exactly twenty-four hours, shorter, or longer. According to the rules of entrainment, each of these body clocks has to expose different portions of its response characteristics to the light (hiding the others in the dark). Individuals whose internal time is controlled by these clocks would each have a different chronotype. Besides the fact that every clock produces a distinct chronotype (depending on its internal day length), there is an important difference between those clocks that produce internal days shorter than twenty-four hours and those that produce internal days longer than twenty-four hours. When the strength of the zeitgeber decreases, the former become even earlier and the latter become even later. Or conversely, when the strength of the zeitgeber increases, the former become later and the latter earlier.

Joseph grew up on a farm and continued to work predominantly outdoors, being exposed to a strong zeitgeber. When Frederic started to work in the factory, he remained predominantly indoors, and since his body clock produced an internal day longer than twenty-four hours (as in the case of most humans), he consequently became a later chronotype. The main consequence of industrialization is that people work in buildings and that more and more people live in densely populated settlements. The difference in light intensities within buildings and outdoors is enormous. While intensities in a well-lit room hardly ever exceed a couple of hundred Lux, intensities outdoors reach 10,000 Lux even on the rainiest of days and about 150,000 Lux on cloudless days.[2] But not only do town people get less light during the day (because they are mostly indoors), they are also exposed to more light during the night because cities are more brightly lit than rural regions. Thus, the amplitude of the light–dark cycle is much smaller in towns than in the countryside.[3] The stronger a zeitgeber, the more efficiently it compresses or expands the internal day.[4] Consequently, different zeitgeber strengths also create different response characteristics so that the clock has to change the rules of its hide-and-seek game. The top panel in the graph on the next page may look familiar from our discussion of days on other planets. It represents a body clock that produces internal days longer than twenty-four hours.

Since a stronger zeitgeber will increase the compression area more than the expansion area, the clock has to hide more of the former in darkness. As a consequence, we become earlier chronotypes when we are exposed to more light during the day (and to less light during the night). Inversely, if the zeitgeber strength becomes weaker, individuals become later chronotypes, as Frederic did when he started to work predominantly indoors.

You probably have experienced how much earlier you can fall asleep after having spent a day in the country, at the beach, or in the mountains. A German proverb says, "Fresh air makes you tired."[5] Greater physical activity and fresh air can certainly contribute to ex-

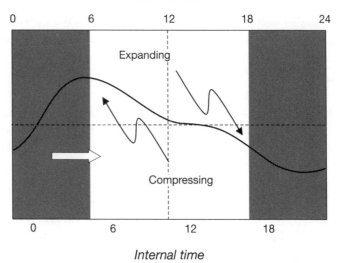

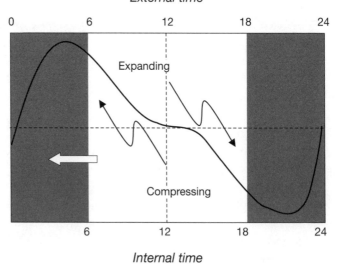

Above, a body clock that produces internal days longer than twenty-four hours. A stronger zeitgeber will increase the compression area more than the expansion area *(below)*.

haustion, but being utterly exhausted didn't necessarily help Sergeant Stein to fall asleep promptly. One reason why we can fall sleep earlier after having spent a day in the "fresh air" is presumably because we were exposed to brighter light—even on a rainy day—and bright light will more efficiently compress the internal day and thus make us earlier.

So far, we have considered the situation of a body clock that produces internal days longer than twenty-four hours. The rule for these clocks is: the brighter the light during the day, the earlier the chronotype. But why then did Hanna go to bed earlier and not later when she stopped leaving the farmhouse during her depression? She is an extreme early type whose internal days are shorter than twenty-four hours. She therefore became even earlier when her zeitgeber became weaker. Joseph, Frederic, and Sophie intuited correctly that Hanna's sleep habits would improve if she helped more actively on the farm and thus spent more time outdoors. More zeitgeber strength would allow Hanna to stay up long after 6 P.M.

In the preceding chapter, you read about the east–west gradient of chronotypes that almost perfectly matches the progression of the sun. However, I have not shown you all the results of this study. The systematic east–west gradient in Germany shown in that chapter was composed only of locations with up to 300,000 inhabitants. The graph on the following page shows that in cities with a population of up to half a million (squares in the graph), chronotypes are later and the slope is flatter than the migration of sun time from east to west (dashed line). The average chronotype of people who live in even bigger cities (more than half a million inhabitants—shown as triangles in the graph) tends to be even later, and the slope is even flatter, even though the systematic east–west gradient is still significant.

This chapter helps you to understand this phenomenon. People in rural regions are exposed to a stronger zeitgeber compared with those living in big cities and are therefore earlier as well as closer to the progression of sun time from east to west. You can even see this phenomenon in regions with less than 300,000 inhabitants (the gray circles). The slope in the east (right half of the gray circles) is steeper than in the west (left half of the gray circles). This is true not only for the whole of Germany (including the states of the former GDR) but also for the southern states of Baden-Wuerttemberg and Bavaria. A much better explanation than culture for the different slopes is population density: its increase from east to west is steeper in the eastern half of the country than in the more populous west.

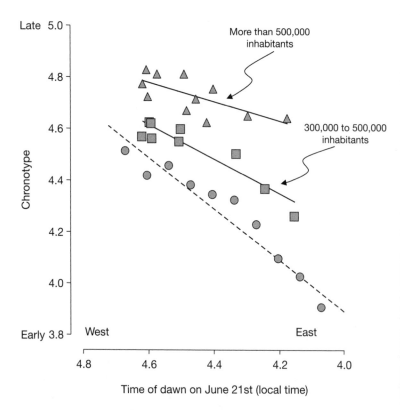

The influence of light as zeitgeber for the body clock depends on population size. The larger the city, the later the chronotype. But even big cities still show a significant east–west gradient.

One of the questions in the Munich ChronoType Questionnaire asks participants to indicate how much time they spend outside every day, without a roof above their head. It is remarkable how little time many people spend outside of buildings or some kind of vehicle. According to our survey, about half of the Central European population spends on average less than an hour outdoors on work days and less than three hours on weekends. When we correlated the time people spend outdoors with their chronotype we found a very systematic relationship that can well explain the differences in sleep–wake behavior between town and country folk.

Based on these highly subjective self-assessments, spending two hours or more outdoors can advance an individual's chronotype by about one hour. This means that if we were to cycle to work and back instead of taking the subway, we might become one hour less sleep-deprived every night during the workweek and would therefore have less sleep loss to compensate for on our free days. And we would also boost our learning capacity, our immune system, our mood, and our social skills.[6]

People in industrialized settings are exposed to zeitgebers about 200 times weaker than those synchronizing the clocks of people who work outdoors. Our modern life in constant twilight has substantial consequences for the timing of our physiology. We can only guess the consequences on our health and well-being of this—on an evolutionary scale—relatively recent lifestyle.

You have seen how different zeitgeber strengths affected the twins and their wives. The solution the quartet found to reinstate their happiness went much further than merely reorganizing the centers of gravity of the women's occupations. I am sure you have guessed that the rearrangements also extended to their emotional centers of gravity. The four swapped partners and thereby ensured that each couple spent a lot of quality time together, both during the day and at night. The new arrangement was socially awkward among friends and family at first, to be sure. But in the end the new couples' happiness was apparent to all. Hanna was able to enjoy more of the evenings together with the other three in the big farmhouse kitchen. Only Sophie became an even later chronotype after she stopped working outside (she eventually took over Hanna's parents' shop). But she did thoroughly enjoy her lonesome nights in the kitchen looking over the valley bathed in magical light around the full moon.

19

Edgar Mass made little notes in his small notebook. He had been up before the sun to measure all aspects of dawn right up to actual sunrise.[1] He stopped making measurements when exactly half of the sun's disc had risen above the horizon. He added some last entries to the long lists and tables he kept in his little black notebook. His handwriting was tiny but immaculately clear. He packed away the small devices he always carried with him to track the sun during the day and the stars at night: a tiny monocular telescope of excellent quality; a small light meter; an altitude meter; and a pocket sextant. When everything was neatly stowed in a small pouch, he went downstairs for breakfast.

Edgar had retired early from his job as a Measurement and Control Technology (MCT) technician but basically continued what he had done all his life: he measured. During his professional life he had developed a deep distrust of what other people measured. He only believed his own measurements. He was obsessed with measuring the world—every aspect of it. His heroes were the naturalists and explorers of the nineteenth century who had shown that the only path to discovery and understanding of the laws of nature was by exact quantification. All modern knowledge, he thought, had been achieved only by the exact measuring of everything.

When he came down, Ingrid had prepared breakfast, but as usual she left the eggs to Edgar. For him, cooking the perfect breakfast egg was a ritual that considered height above sea level, air pressure, and other influences. "Good morning, Ingrid," he said, entering the kitchen. "Sun came up at the right time again."

"Fascinating, dear," she replied, and gave him a flash of a smile without directly looking at him.

"You know how comforting it is to see the sun always come up at exactly the time it is supposed to?" he continued. "183 seconds earlier than yesterday."

"How interesting, Edgy dear," she repeated, and flashed her broad smile again.

"Since we have to leave soon, I think we should skip the eggs today," Edgar said.

Ingrid replied, "As you wish, Edgy."

They would be leaving the house for the airport in less than an hour to catch a plane for Casablanca, from whence they would continue their journey by bus to southern Morocco. Ever since Edgar had retired, they spent their summers in a small apartment that had a roof terrace overlooking the ocean. Edgar found this spot ideal for measuring the movements of the stars, the planets, the moon, and the earth. The accuracy of the celestial players gave him a deep feeling of security. He always chose the day of their annual departure to fall just before Europe switched the clocks to daylight saving time. Edgy detested the human hubris of messing with the sun's time. He always felt like a proud fugitive, who left his country like those who had to leave for political reasons.

A couple of hours later, they were crammed into their economy-class seats, trying to master the almost impossible task of juggling the multitude of objects on their miniature foldable trays: containers filled with typical airplane food; plastic glasses containing water and juice; two cups of coffee; Ingrid's crossword puzzle; and several of Edgy's measuring devices, one of which was an altitude meter. The captain had just turned on the intercom and was providing the passengers with crucial information about the weather, their estimated flight time, and the local time at the destination. "We are cruising at an altitude of 30,000 feet," he said, concluding his address to the cabin. "Enjoy your meal."

Edgar looked at his altimeter and frowned. He turned to Ingrid. "But that's not correct, our altitude is 29,700 feet," he said. "This is bad; if their devices are incorrect, we might crash into another plane."

"Are you sure?" Ingrid looked worried, but Edgar gave her a withering glance, as if to say: "Have I ever been wrong?" Edgar pressed the call button, and when the stewardess arrived he told her that he needed to speak to the captain on a matter of life and death.

The first response of the stewardess was panic. But looking at Edgar she decided that he was an unlikely candidate for a terrorist, so said, "I'll see what I can do." A couple of minutes later the co-pilot came from the cockpit, and Edgar told him that their altimeters were obviously not working properly. He was very relieved when he was told that the captain's declaration was an approximation and that Edgar's measurements were, of course, right on target. He needn't worry, the co-pilot reassured him, because 29,700 feet was exactly the height the plane should be flying at. Ingrid pretended to be deep into her crossword puzzle during the entire incident.

The rest of the trip was uneventful, except for the little incident at the currency exchange counter, where Edgar challenged the conversion rate. They arrived at their summer home late in the evening and Edgar immediately went up to the roof terrace to take some measurements before retiring. After a short night of sleep, he was back up there again before dawn. While finally packing away his instruments into his small pouch, he heard someone come up the stairs leading to the rooftop. He was slightly confused because Ingrid normally would still be fast asleep at this time of day during their African sojourn. Still bending over his pouch with his back toward the opening where the stairs ended on the rooftop, he greeted her in his usual way. "Good morning, Ingrid, sun came up at the right time—at least here in Morocco." Since she didn't answer, he continued, "Same local time as back home, by the way, which means dawn was later than yesterday! It's spring, the sun should come up earlier every day, not later—it's all so wrong!" He turned around, anticipating Ingrid's usual reply: "Fascinating, dear!" Instead she just stood there quietly and, to his surprise, was looking directly in his eyes while producing her usual flash of a smile.

They found Edgar's body later that morning on the pavement in front of the house. The police had to wake Ingrid, who was still fast asleep when they rang the bell. The inspector came to the conclusion that Edgar must have tripped over his pouch, and so slipped over the rooftop's edge while looking through his telescope, which was lying not far from his body. All his other instruments were scattered around him, and the empty pouch had, by chance, ended up covering his distorted face.

Once all the formalities had been settled, Ingrid sold the apartment, returned to Frankfurt, and never went back to Morocco. Every year when daylight saving time came around, however, Ingrid went to church and lit a candle for Edgy, displaying her famous flash of a smile.

Obsessed people like Edgar are convinced that nature is the only reliable source of information in today's world of constant make-believe and blatant lies. It is therefore not surprising that they cannot tolerate how society so profoundly messes with the natural course of our lives—for example, when it changes our social clocks twice a year. The population seems pretty much torn between two camps when it comes to daylight saving time (DST)—one considering it a vice, the other a virtue. According to the latest (nonrepresentative) internet polls of spring 2009, the anti-DST faction is gradually growing. DST supporters remind their foes that clocks are changed by *one small hour* and in the right direction, so that the change supports the natural seasonal progression of the sun. But few studies have properly investigated whether people, or rather their body clocks, adjust to these time changes, and if so, how long this adaptation takes. Most of these studies have looked only at the first week, haven't separated work days and free days in their analyses, and also have not considered internal time (chronotype).

Using Germany as an example, I have described how tightly our body clock is tied to the natural light–dark cycle. Despite the overwhelming impact of social schedules on our lives, which are more or less similar throughout any given country, the body clock becomes later from east to west by approximately four minutes per degree of longitude—the same amount of time per degree of longitude that the sun takes to move from east to west.[2]

In view of these results, we were curious to know how body clocks coped with the artificial transitions of our social time, namely when we move in and out of DST. We therefore recruited volunteers willing to participate in a study that accompanied individuals for eight weeks around the transitions from and to DST. The study started with the autumn transition from DST to normal zonetime, followed by the inverse transition in the subsequent spring. The volunteers received a fat envelope via mail one week before the study started. The package contained several questionnaires, which had to be filled in before the study started, as well as sleep logs, which they had to fill in every morning. The sleep logs asked questions about their past twenty-four hours.[3] The package also contained an actimeter that had to be worn throughout the eight weeks of the experiment.

In addition to the field study, we investigated how the general population copes with DST transitions and how their sleep–wake habits change over the course of the year. At that time, we had over 55,000 entries in our database—all with a date. To analyze seasonal changes in circadian behavior, we average the wake-up times on free days for each half-month. These are plotted downward, progressing through the year, beginning with December at the top. The same series of twenty-four data points is plotted twice, one below the other, to help us see the changes across the winter months. The horizontal axis represents the time of day from 5:00 to 9:30 in the morning, based on Standard European Time (SET). The edge of the gray area indicates sunrise.

It is quite remarkable how tightly daily human behavior is coupled to dawn on free days. Following sunrise, people are on average getting up earlier from midwinter into the spring, as we had expected from our east–west study across Germany. What we hadn't foreseen is that this dawn-tracking stops as soon as Central Europe switches to DST on the last Sunday in March.[4] Throughout DST, wake-up times on free days scatter around a stable social time, around 8:45 A.M. (corresponding to 7:45 SET). Yet, when the social clocks are switched back to the "normal" zonetime on the last Sunday in October, dawn-tracking of the wake-up times on free days immediately reappears. Thus, the body clock adapts its phase of entrainment (chronotype) to season. But why does it do so only during the non-DST months of the year?

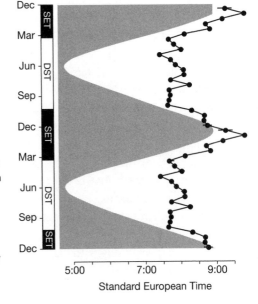

An average of the wake-up times on free days for each half-month from the Munich ChronoType Questionnaire database. At the time of the study, the database contained 55,000 entries. The right border of the gray area represents dawn in Central Europe.

It is unlikely that the body clock would actually track dawn throughout the entire summer—even without DST. Think of people who live close to the poles, where days may never end in midsummer. In addition, no records from the pre-DST era suggest that they tracked dawn throughout summer. Since they would wake up at around 5 A.M., they would have to be asleep at between 9 P.M. and 10 P.M., unless they would severely shorten their sleep.

However, the fact that the body clock drops and picks up tracking dawn exactly at the DST transitions indicates that these artificial time changes may well have an influence on our body clock's natural adjustment to season. The transition to DST is scheduled around the spring equinox, while the release back to SET is usually more than a month later than the autumn equinox.[5] These two asymmetrical transitions correspond to quite different day lengths, yet they still elicit the same prompt response in our sleep–wake behavior. If body clocks were to stop tracking dawn at a certain day length in spring, we would expect that they would pick up to do so at the same day length in autumn.

According to the results from the large database, arguably the DST transitions merely supported a natural tendency in spring to wake up earlier and in autumn to wake up later, and that the body clock should therefore adjust quite easily to these social time changes. However, the results of our field study indicate that this is not so. The sleep–wake behavior in the relatively small group of volunteers is very similar to that found in the large database. During the four weeks before social clocks were switched from DST back to SET in October, they woke up (on free days) at approximately the same social time—their body clocks did not track dawn. Once the DST transition occurred, however, wake-up times immediately went back to tracking dawn.

At first inspection, the sleep–wake behavior of our volunteers around the spring transition indicated that body clocks easily adapt to DST. Yet, when we separated the results according to chronotype, it became clear that only early types adjusted perfectly in their sleep–

wake patterns. Late types had still not fully adapted by the fourth week after the time change. When we then analyzed the volunteers' activity–rest rhythms recorded by the actimeters, the inadequate adaptation became even more apparent.

The daily activity recordings of different chronotypes mirror their respective sleep–wake behaviors. The activity profiles (on free days) shown in the next graph are averaged over the four weekends before the DST change in autumn. The typical free-day activity profile of an early chronotype is drawn in front and that of a late type in the back. While the former begins to be active at around six, the latter doesn't start before 10:30. At the other end of the day, the two profiles show approximately the same time difference. In this specific example, judged by the times of inactivity, the late type appears to be a slightly longer sleeper than the early type.[6]

The next graph illustrates the weekly averages of activity onsets (on free days) of early and late chronotypes and shows that they par-

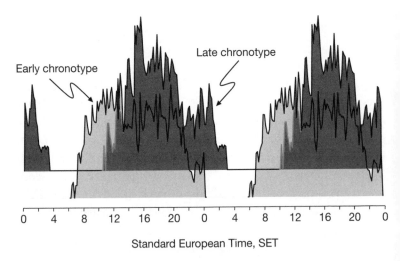

Standard European Time, SET

The free-day activity profiles of an early chronotype *(front)* and a late chronotype *(back)*, averaged over the four weekends before the daylight saving time change in autumn.

allel dawn (the right borders of the gray areas) before the DST transition in the spring.[7] Onsets are a bit scattered for the early types but almost form a line parallel to dawn for the late types. At the DST transition, early types jump one hour in their onset times while late types do not seem to respond at all. Yet four weeks into DST, neither of the two groups had fully adjusted to the social time advance of one hour. The early types had shifted only by approximately forty-five minutes, and the late types went back to where they had started eight weeks before.

But why do we have such difficulties in adjusting our body clocks

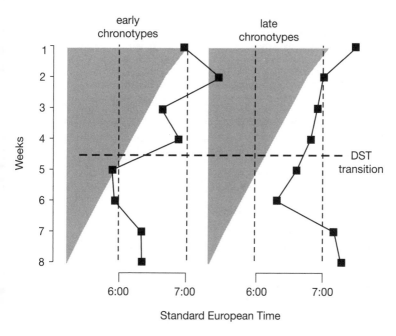

Weekly averages of the calculated maximum activity times on free days of early and late chronotypes. Both parallel dawn (the right border of the gray areas) before the DST transition in the spring. Early types have not advanced their activity by a full hour even after four weeks, while late types appear not to adapt to the clock change at all.

to the regular time changes when moving in and out of DST? After all, it is only *one small hour!* The answer becomes quite obvious when we look at what happens at these time changes in more detail. When we emerge from the long winter nights, we obviously adjust our body clocks to the gradually advancing dawn times. This adjustment is only apparent on free days. On the more frequent work days, we get up more or less at the same local time—in most cases earlier than on free days. At first we get up before the sun, then with her, and then after her rise. This continues for approximately three months until the clocks are switched to DST in March.[8] From one day to the next, we are thrown back by approximately three weeks and now get up again before the sun rises. The closer the DST change to the equinox, the shorter is the time we are thrown back in our natural seasonal progression. In the autumn, this scenario is reversed. With the days getting shorter again, we first get up after the sun, then with her, and then before dawn. When the clocks are set back (in Europe approximately a month after the equinox in October), we are thrown back by approximately four weeks (the exact amount depends on latitude). Follow the thick black line (sunrise) in the next graph from left to right (a hypothetical wake-up time of 7 A.M. is indicated by the dashed horizontal line). The spring time-change that throws the seasonal progression back by three weeks (if one lived like Ingrid and Edgar in Frankfurt) is equivalent to travelling fifteen degrees of longitude to the west; the autumn DST transition, which throws our seasonal progression back even more, reverses this virtual travel, transporting us back fifteen degrees to the east. In addition to the abrupt time changes, the seasonal amplitude of dawn is greatly reduced in relation to our daily work life. (The white dotted curve indicates the progression of dawn without DST.) Reducing the sun's seasonal amplitude is equivalent to a virtual journey toward the equator. For people living in Central Europe, the interference of DST therefore corresponds to a virtual journey from Frankfurt to the south of Morocco and back.

In reality, DST is nothing but a collective decision to start work an hour earlier or to work for a company situated one time zone farther to the east, without ever leaving our hometown (similar to what Timothy did in working for LayIn&Out). A cynic might argue that we are made to change our clocks only so that we don't notice this collective decision, making us believe that we go to work at the same time. Since the results of our field study indicate that body clocks do not properly adjust to the time change (especially those of the later chronotypes), this social intervention is bound to increase our social jet lag and decrease our sleep duration.[9]

The broad smile that Ingrid produced at every DST change while lighting a candle for Edgy did not only express her relief at no longer having to cope with his measuring mania. She had also long realized the irony that lay in their annual exodus to southern Morocco. Didn't it essentially amount to the same changes between their daily behavior and sunrise created by the DST transitions that Edgy abhorred? They might just as well have stayed in Frankfurt.

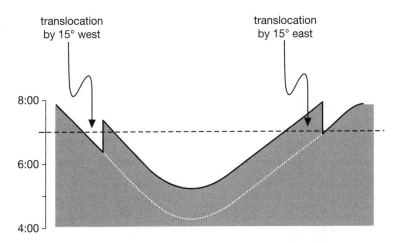

Daylight saving time drastically affects the relationship between sun time and our daily habits, similar to traveling southeast in spring and northwest in autumn.

20

Marco Gonzales took the call and looked at his computer screen that showed him all the details of the weather in Manchester. "Good morning, Mrs. Taylor. How are you today? Isn't it great that it finally stopped raining?" Marco had just started work and Mrs. Taylor's call to the bank was the first of an endless series he would be taking for the next twelve hours. "My name is Marco. How can I help you?" Mrs. Taylor asked him to pay a doctor's bill from her account. She was elderly and handicapped and wouldn't have known how to bank online even if she had owned a computer. "Certainly, that's what we are here for," answered Marco cheerfully. "Why don't you first give me your PIN and then the name of the doctor and his or her account details." After typing the information into the mask on his computer screen, he said, "and now I will need an identification number for this transaction, Mrs. Taylor." She slowly spelled out the six-digit number. "I am sorry, Mrs. Taylor, but this number has already been used. Why don't you take the next one on your list?"

For the next three hours, Marco dealt with about thirty more customers. Between calls, he rolled his chair back to cast a glance at Maria, who worked three stations away in an endless line of cubicles. It was a particularly busy day and he had had no luck in making eye contact with her. Whenever he tried, she was concentrating on a client. They had met six months ago during the introductory training program that had taught them all the necessary skills for their job, including communication skills and intensive language coaching.

Besides several big rooms crammed with rows of three-walled

cubicles, each of which contained only a computer and screen on a tiny desk, the facility included several small rooms with two bunk beds each. Employees could take a nap in these whenever they felt tired. The frequency and length of such recuperations was strictly regulated. Marco took off his headphones and stood up to look into the row of cubicles parallel to his. Cubicles A04 and A11, normally occupied by Roberto and Miguel, were empty. They both must have gone to take a nap or have a smoke outside. Marco noticed that they had recently synchronized their breaks and wondered why. Today was Miguel's birthday, and a dozen or so of their friends had arranged to meet in a bar after work to celebrate. Marco sat down again and rearranged his headphones to take a call. Having dealt with the customer, who was complaining about an erroneous transaction, he rolled back his chair again and finally caught Maria between two calls. She routinely turned around after finishing with a client to see if Marco was looking at her, and her face lit up when she finally managed to make contact. Using their usual sign language, they arranged to meet in their accustomed bunkroom in twenty minutes.

When Maria and Marco were finally done for the day, they walked out of the building and were dazzled by the bright day outside. Their eyes took quite some time to adjust. Most of the rooms of the facility had no windows, so when they emerged they often felt like miners coming back to the earth's surface. They had arranged to meet the others near the hangout area, where they usually went during their breaks for strong coffee and cigarettes. When everyone had gathered, they walked for less than a mile to their favorite bar. While walking to the bar, Marco turned to Maria and said, "Did you hear a typhoon is on its way?"

Maria looked at him, slightly worried. "Oh no, not another one. I hate typhoons, and we have had enough already this season."

Marco put his arm around her. "Maybe it will turn another way. We'll be all right, you'll see."

The bar had designed itself around the needs of the young workforce of the newly developed industrial park. There were no neighbors to complain about the noise that surrounded the place at the oddest hours. The party went on for almost five hours and everyone had a lot to drink. Fortunately, they all had the next three days off work. Everyone was in an excellent mood when the party began, but the atmosphere went downhill after José, who was quite drunk already, started to report the latest gossip about who had broken up with whom and couldn't stop himself revealing rather confidential information about some of the new liaisons that had been the causes of the break-ups. His insensitive disclosures sparked several rows and a fair bit of shouting; in the end, three women were in tears and five men left the party in a sulking rage. It was already around lunchtime by the time the party was finally over, but Maria and Marco didn't want to let each other go. Like most of their friends, they still lived at home with their parents, and so they would not see each other for three endless days. They and all of their friends loved their work, and not only for the good salary. It also gave them the opportunity to create their own subculture, away from the strict rules of their families.

One of the most blatant assaults on the body clock in modern society is shift work. At present, close to 20 percent of the workforce in the industrialized world is engaged in schedules that deviate from traditional work hours and, in most cases, these involve rotations. Several decades of epidemiological research have clearly shown that shift workers develop more health problems than day workers. These include sleep problems, depression, cardiovascular pathologies, digestive tract issues, diabetes and other metabolic diseases, and obesity. Health risks related to shift work even include several types of cancer. As a consequence, the World Health Organization has recently classified "shift work that involves circadian disruption" as a potential cause of cancer. In 2009, the state of Denmark started to pay compensation to women who have worked in night shifts and subsequently developed breast cancer. In the Netherlands, shift workers have started to file legal suits for compensation for various health problems potentially linked to their long-term shift rotations.

There is no doubt that shift work is detrimental to health, but the mechanisms linking shift work and pathology are far from well understood. The issue is complex. Shift work means eating when the body is not optimally adjusted to digest food and not eating when the stomach is prepared to do so. It means sleeping outside of the sleep window provided by the body clock and working during our internal night. It means being exposed to light when the eyes "expect" darkness. It means not being around when a partner and children end their day or when they get up in the morning. It means trying to sleep while the children want to be active, when the streets are full of noise, and when it's bright daylight outside. It means constant disruption of one's social life and being exhausted even on free days. (On free days, shift workers try to fall back into synchrony with family and friends.) It means stress for body and brain, leading to coping and compensation attempts—for example, by consuming large amounts of caffeine or by smoking many cigarettes, as we know people do when they suffer from social jet lag.

Initially, shift work was mostly confined to professions responsible for our health and safety (hospitals, police, fire brigade, and so on). Then it spread to the production industry, where expensive machinery needed to be utilized around the clock. More recently, shift work has reached service providers like the call center described here. One would think that the call center of a bank would keep relatively normal working hours, maybe with extensions outside of the traditional nine-to-five arrangement. After all, few customers want to make transactions at 4 A.M. The reason for shift work in the service sector is outsourcing. Millions of people in India, the Philippines, and other Asian countries work in the Business Process Outsourcing (BPO) industry. In the metropolitan region of Manila alone, more than 240,000 people work in call centers and other BPO facilities. When a British bank outsources its telephone services to another country, the service still has to be provided during British working hours. Nine to five in Britain means 4 P.M. to midnight in Manila; and nine to five in Boston is exactly twelve hours later (or earlier) there. The employees of an outsourced call center have to pretend to live in the same time zone as their customers, and so they have to speak the way their customers do, and they have to know about the local weather at their client's location.

Shift work means living at odds with a normal social life as well as against the body clock. It is therefore not surprising that it harms health. Although shift work may be the immediate cause of certain illnesses (for example, ulcers, because of eating at the wrong times), it is unlikely that it is the direct cause for all the pathologies that epidemiological studies have shown to be associated with shift work. A more likely scenario is that the constant physiological stress weakens the system and thereby also weakens its ability to keep disease at bay.

The *light-at-night (LAN) hypothesis* proposes that shift work can directly cause tumor development. This hypothesis follows a chain of six arguments, each of which are correct as individual statements: (1) melatonin is a hormone produced under the control of the circadian clock at night; (2) melatonin synthesis can be suppressed by light; (3) melatonin belongs to the chemical substance class of indolamines;[1] (4) indolamines can act as scavengers of oxygen radicals; (5) oxygen radicals can damage DNA; (6) DNA damage can cause cancer.[2] The idea behind the LAN hypothesis is based on the observation that the production of melatonin is inhibited by light, and since shift workers are exposed to light at times when their body normally produces melatonin, less of this indolamine will be circulating in their blood compared with people who sleep in darkness at the right internal time. The presumed reduction of melatonin in shift workers is therefore thought to cause cancer via increased DNA damage by oxygen radicals.[3]

Advocates of the LAN hypothesis argue that the potential of melatonin to reduce cancerous growth has been shown in animal experiments. However, these experiments rely either on cancer cells already being present, or on such cells being purposely injected into the animals. Thus, if someone had already developed a tumor, shift work could certainly promote further cancerous growth. But that doesn't mean that shift work actually caused the cancer in the first place. Thus, we are back to the more likely scenario; the constant physiological stress of shift workers weakens the system's ability to keep potential pathological developments at bay.

The LAN advocates argue that epidemiological studies have shown cancer rates to correlate with the degree of light pollution. But light pollution is also a clear indicator for an industrialized lifestyle, which includes many factors that may individually or in combination contribute to increased cancer rates. Thus LAN is more likely to be a cancer predictor than a direct cause for cancer. The media frequently pick up on the proposed connection between LAN and cancer. Some media reports have even insinuated that *any* light pollution—from streetlamps to nightlights—could lead to the development of cancer. But even if reduced melatonin levels promoted tumor growth, it is unlikely that melatonin is suppressed in people who are active (and exposed to bright light) during the day and sleep in a relatively dark room during the appropriate internal times. For light to be effective in melatonin suppression, one would have to assume that people sleep with their eyes open in a relatively well-lit room.

Under normal circumstances, closed eyelids reduce the amount of light reaching the retina by more than 80 percent. Our reddish eyelids filter away even more (97 percent) of those wavelengths that are most efficient in suppressing melatonin. A lamp producing, for example, 100 Lux shining directly on closed eyelids would thus expose the retina to no more than three Lux. This is less than a third of the intensity that researchers use in controlled laboratory studies investigating the body clock when they want to be sure not to affect melatonin levels or other physiological parameters. But even the three Lux are an overestimation because our eyeballs roll upward during deep sleep, reducing retinal exposure even further. It is therefore extremely unlikely that melatonin is suppressed during our sleep by nightlights, televisions, computers, emergency lights, moonlight, streetlights, or any other urban light pollution. Even if we slept in a bedroom during the day with closed curtains (for example, following a night shift), melatonin should not be affected. Under these circumstances, it is unlikely that light pollution can be directly linked to health via the LAN hypothesis. Instead of being afraid of light at night, we should pay more attention to the fact that we, as industrialized beings, expose ourselves to far too *little* light during the day. It is more likely that the weak zeitgebers in urban settings—due to relatively dim daylight exposure in combination with more light at night—have far greater effects on our mood and on the adequate synchronization of our body clock.

One major factor in the context of the LAN hypothesis or in shift-work research in general is almost never considered: internal time. Practically all shift-work studies analyze their data based on external time. For example, night shifts in these studies start at 10 P.M., morning shifts at 6 A.M., and evening shifts at 2 P.M. Yet, the reference for all daily aspects of our physiology is individual *internal* time. Let's pretend that a study wants to find out whether earlobe cancer develops more readily in shift workers than in day workers.[4] The study collects epidemiological data, analyzes them, and finds no difference in the development of this pathology between shift and day workers. If, however, the scientists had assessed their subjects' chronotype, they might have found a different result. Early chronotypes develop earlobe cancer with a higher probability if they work at night while late types have an even lower risk in contracting this pathology if they work at night. This fictional example indicates that, without the consideration of internal time, we will not fully understand the relationship between shift workers and pathologies.

In view of the lateness of most people in our modern societies, one could argue that a majority of the workforce is scheduled in a permanent early shift when they work from nine to five. The diversity of chronotypes—with extreme larks and extreme owls being up to twelve hours out of phase—predicts that some chronotypes may even benefit from working at night and others would be more in synchrony with their body clock if their working days started at four in the morning. People working in rotational shifts are generally younger than the average worker. One could argue that the reason for this bias lies in the higher capacity of younger people to cope with the physiological stress of rotational shift work. Yet people become earlier and earlier in their chronotype with age and thereby may lose their endurance and their capacity to stay up late.

All the protagonists in this chapter's story are just beyond the developmental peak of lateness and therefore can work at night more easily than older people. They can also sleep longer and better during the day after getting home. Note that this is not due to the fact that their body clocks adjust to the constant night shifts like the oilrig workers you read about. When Maria and Marco stepped out of their workplace after finishing their night shift, "they were dazzled by the bright day outside." The light detection system of the body clock integrates light over time.[5] Even if the workplace were well lit, the eyes would probably not be exposed to more than 100 Lux. In a twelve-hour shift, they would therefore integrate 1,200 Lux-hours. Let's assume that the outside light intensity on their way to the bar was 120,000 Lux and that it took them twenty minutes to walk the short mile. In that case, they would have been exposed to 40,000 Lux-hours. But even in the rainy season, with only black clouds in the sky, they would still have accumulated 3,000 Lux-hours on their way to the bar. As a result, their body clocks don't fully adjust to the nocturnal work hours. The reason they can cope better than older people is that their body clocks are extremely late, even if they remain synchronized to the normal daylight.

Besides the medical difficulties that are associated with shift work, there are many social issues: fathers and mothers who are out of synch with their children and their partners, as well as with their friends and relatives. As this chapter's story described, shift workers often form subpopulations, in this case made up of youngsters who work at the same place and during the same odd hours. These subpopulations rarely mix with the rest of society. This form of social isolation has even led to the observation that the number of HIV infections in India and other Asian countries is higher among employees of call centers than in the general, age-matched population.

21

Louise was dead tired but just couldn't find sleep. Normally she didn't mind that Bruno was still reading beside her. But tonight he seemed to flip his pages especially loudly, and his bedside lamp appeared brighter than usual. After a lot of tossing and turning, of rearranging her duvet, of sighing deeply and frequently, she opened her eyes and turned to her husband. "Bruno," she said in a suffering voice, "would you mind not reading tonight? I just can't fall asleep."

Now Bruno sighed deeply. "All right, I'll sleep in the guest room," he replied. He took his book and his cushion and swung the duvet over his shoulders, thereby knocking over his bedside lamp. It made a lot of noise but luckily didn't break. He ignored Louise's angry glower. "Serves her right for shoving me out of my own bed," he thought. "She's the one who can't sleep, so why didn't it cross her mind to retreat to the guest bed herself?"

Louise and Bruno had been married for twenty-eight years, and their children had all started their own lives outside of the parental nest. At last they had enough room for "self-actualization," as it is called. One of the children's bedrooms was turned into a guest room, the other into Louise's parlor, and the third into Bruno's den—at least that's what the children ironically called the converted rooms. Since neither parlor nor den contained a bed, Bruno often evacuated to the guest room. Despite his grumpiness about being shoved out of his own bed, he quite liked to sleep in the guest room. It meant that he could read as long as he liked and could even listen to the radio or watch television whenever he felt like it. But even more important, he could sleep late in peace. Louise usually got up at the crack of dawn (at least in his slightly exaggerated opinion) and then opened the curtains to "check the weather." That was the most annoying part of her morning ritual. It was she who insisted that the curtains were drawn in the first place, to prevent her waking up even earlier. He quite liked to sleep with the night's air and the morning light coming into the room—he could sleep well into the day after sunrise. It was the sound of the opening curtains that woke him up. If it hadn't been for the subconscious feeling of loneliness during the night, he would have taken his self-actualization a step further and turned the guest room into his own bedroom.

When he came down the next morning, he found a note from Louise on the kitchen table, announcing that she had gone off to have coffee with her best friend, Doris, as she often did on Mondays. He made himself a proper breakfast consisting of eggs and bacon on toast, which he only dared to cook when Louise was away. She had turned vegetarian after their children had left the house, while he still considered himself a true carnivore. After he had finished the last morsels, he sat on the terrace finishing his coffee and thinking about how Louise and he had become so different over the years. They had always been different, of course—it was what attracted them in the first place; but now after almost three decades of marriage it seemed that their initial differences were intensified. It had been a shallow slope from being attractively different, to occupying different and quite useful niches in their partnership, to being so different in many ways that it sometimes became a burden. It seemed to him, though, that they had already lived through the peak of discrepancies.

Before he had retired and their youngest child still (occasionally) lived at home, he sometimes thought that he and Louise lived in parallel universes. They appeared not to share a single common interest—not a book, not a radio or television program. Even their interpretation of newspaper articles went in entirely different directions. But now that he was home most of the time, it seemed that they had started to share more aspects of their lives again. They spent more time with each other and, above all, had more time to talk. In these conversations they rediscovered their common ground, even though many of their widening differences still persisted. When they sat in the car together, he had the impression that she drove more and more slowly every week, while he still liked to be an "agile member of traffic," as he phrased it. For years her driving style had driven him bonkers, but it had become ever less important for him. In most cases, he was now even content to let her do the driving.

While he was enjoying his third cup of coffee and the newspaper on the terrace, he heard their car turn into the driveway. A couple of minutes later Louise came into the house and dumped several heavy-looking packages on the empty chairs around the garden table. "Looks like you've done some serious shopping with Doris," he greeted her in a friendly tone.

"Sure did," she replied cheerfully. "I had a bad conscience for kicking you out of bed last night and told Doris about our difficulties." On the one hand Bruno liked the fact that Louise sort of apologized for her behavior. But on the other, he always hated the fact that Louise spoke about their private affairs with other people. But before either of these feelings could dominate his mood, Louise continued happily, "Doris laughed when I told her, because she and Sean have had exactly the same problems—but they found an ingenious solution many years ago." She opened one of the bags and pulled out a large piece of fabric with the same pattern as their bedroom curtains. "They've installed a curtain that can be pulled between the two halves of their bed." Louise held the fabric up, looking exceedingly pleased with herself.

When I give public talks about the body clock, I often start the lecture with a poll of the audience: "If the following situation seems familiar to you, please raise your hand." Then I recount an abbreviated version of the beginning of this chapter's story, and practically every hand in the lecture hall goes up. When I read this chapter's story to a friend, she interrupted me after the first couple of sentences and said, "Have you been spying on us through our bedroom window?" Another friend just laughed: "Exactly like in my marriage, except our roles are reversed." Divergent temporal habits seem to stand as a proxy for the many other differences that develop between partners in a long-term relationship. Triggered by an overwhelming impression that people always seem to judge their partner to be a substantially different chronotype, I wanted to collect facts, and so I turned to our database.

In the first version of the Munich ChronoType Questionnaire, we had additionally asked the subjects to assess their own chronotype and that of their partner based on seven categories.[1] So for investigating partnership timing, we had the following dataset from approximately 50,000 people: age, sex, actual chronotype (calculated according to midsleep on free days), the self-assessed chronotype, and that assessed for the partner.

As you know, the chronotype of teenagers becomes gradually later until it reaches a maximum in lateness at around nineteen in women and twenty-one in men. Across most ages, men are on average later chronotypes than women. These results are shown again in the next graph. It is almost identical to the one you have seen before, except for an additional symbol: two circles connected with a horizontal arrow. They are part of a joke that I make when presenting this graph (to wake up the audience in the dimly lit lecture hall). I claim that we have not only discovered a biological marker for the "end of adolescence," but that we have also found the reason why men marry younger women. After a little pause, I continue: "because then they can have breakfast together" and only then add the extra symbol to the graph. Since we can quite accurately calculate the chronotype of the subject (based on midsleep on free days), we can query the database about how the different chronotypes, on average, assess themselves and their partners (using the seven categories described above).[2] The results for self-assessment are straightforward: the later their midsleep on free days (MSF), the later both women and men assess their chronotype, showing no differences between the sexes, as the top panel of the figure on the next page shows.

The chronotypes of partners, however, are assessed very differently by men and by women (see bottom panel). Men judge the chronotypes of their partners independent of their own chronotype: their partner's chronotype averages around 3 (intermediate type). This contrasts with how women assess their partners: the later their own MSF-based chronotype, the later they judge their partner's. This striking difference may be explained by another finding. Men go to bed at different times (in most cases later) when their partner is not home, indicating that men adapt their sleeping habits to those of

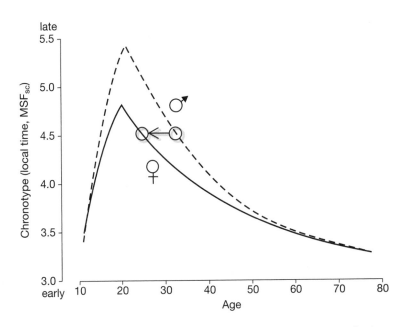

Across most ages, men are on average later chronotypes than women according to our selected sample. The differences decrease as men and women age. Thus, when the man is older than his partner, their chronotypes tend to be more similar.

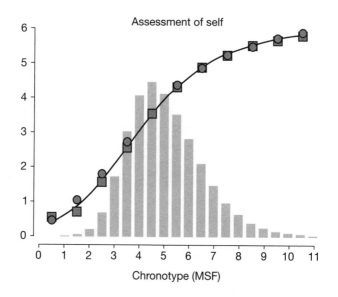

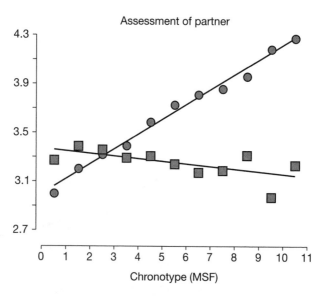

Chronotype assessment of self and others: the later their midsleep on free days (MSF), the later both sexes subjectively assess their chronotype, showing no differences between the sexes (*above;* the gray bars indicate the distribution of chronotypes in the entire database). However, men and women subjectively assess their respective partners very differently *(below).*

their partner. That is possibly why women think that the chronotype of their partner is similar to their own, while men have a more realistic notion of their partner's chronotype. The relationship results in a horizontal line, as one would expect if we don't choose our partners by chronotype.

The next question we approached was whether the way subjects saw their partners influenced how they saw themselves. In an earlier study, we had already shown that the self-assessment of the subjects could be quite inaccurate on an individual basis (compared with their actual MSF-based chronotype) and hypothesized that this could be due to the influence of the partner. For example, if an individual was an intermediate chronotype (based on MSF) but the partner was an extreme owl, then he might consider himself an early type just because the partner's chronotype is much later. Our results confirmed this hypothesis. Individuals who categorized their partner to be an early type assessed themselves to be later chronotypes than they actually were, according to their MSF-based chronotype, while those who judged their partners to be late types saw themselves as earlier.

These results show that chronotype should be assessed by the sleep times and not by subjective self-assessments. There are chronotype questionnaires that ask subjects only about their habits and preferences. The problem is that the answers will always be biased by the habits of other people. The chronotype distribution in India, for example, is much earlier compared with that in Central Europe, which must lead to different self-assessments: a midsleep on free days of 4:30 A.M. would be average in Central Europe but would characterize an extreme late type in India.[3]

The next set of results makes the strong subjective element involved in assessing chronotype even more apparent (and takes us back to Louise and Bruno). When investigating how age affects people's judgment about their own and their partner's chronotypes, we got a surprising result: with increasing age, partners judge each other's chronotypes to be increasingly different, up to an age of approximately fifty-two, when this difference decreases again.[4] This observation is remarkable because, in reality, the actual chronotype differences between men and women continually decrease from twenty to fifty-two (becoming indistinguishable at higher ages). There are at least two possible explanations for this observation, one of which is psychological and the other cultural.

Let's first look at the psychological explanation. Remember Bruno's reflections while having his coffee on the terrace about the "shallow slope from being attractively different, . . . to being so different . . . that it sometimes became a burden." The longer people are together, the more they seem to occupy different niches in their relationship on all levels, ranging from being responsible for different tasks in their daily lives to developing different tastes or opinions. The reasons for this apparent specialization are complex, are not restricted to heterosexual relationships, and also exist in other long-term relationships (siblings or colleagues, for example). One trivial explanation is that people change with age and, in partnerships, not necessarily in the same direction. It is therefore not surprising that partners also gradually view their partners as different chronotypes. The fact that these apparent discrepancies become less severe at higher ages may have many reasons. Decreased sexuality, for example, might weaken certain rivalries between partners and thereby also reduce the need to be different. Another factor is retirement; partners have to rearrange their lives when they stop working.

Statistics show that the number of divorces depends on age. Divorce reaches a maximum when spouses are between forty and forty-five. People generally divorce each other because they no longer have anything in common or because they have found another partner, which sort of amounts to the same thing. The decrease of divorce rates at higher ages has in part a very simple reason. Those people who don't get along with their partners are already divorced by the age of fifty-five and may have found new partners with whom the slow divergence process is only just beginning. This principle could also be behind the decline of subjective chronotype differences. As you can see, I cannot offer solid psychological explanations for our observations on partnership timing and therefore meander through many different possibilities. Scientists call this "hand waving." Fortunately I can discuss the second, the cultural explanation, with more scientific reasoning.

We store individual, one-time entries in our database, which makes our analysis a *cross-sectional* study.[5] These studies have the disadvantage of not being able to distinguish between age-related differences and those that are due to the fact that subjects grew up in a different era. The majority of sixty-year-old subjects who filled out the questionnaire probably got married several decades ago, when partners did not live together prior to their wedding. Since they had no experience in sharing life with a different chronotype, this quality of the partner didn't influence their choice. In contrast, nowadays young people start to live with each other before getting married and might decide that always having breakfast alone or spending most evenings alone while their partner is already asleep speaks against forming a long-term relationship. In that case, we would expect that younger couples are generally more similar in their chronotype than older couples. Our data clearly rule out this possibility: chronotype does not correlate between partners, no matter whether they are young or old. The psychological explanation is therefore more plausible than the cultural: the longer people live together, the more differently they see each other.

Bruno, by the way, rejected the curtain dividing their matrimonial bed and confessed to his wife that he rather enjoyed falling asleep in the guest room. Eventually they reached the very amiable compromise of sometimes sleeping apart and sometimes together.

22

It was 7 P.M. on the twenty-second of September when Gerry's cell phone started to play Pink Floyd's "Time" from the album *Dark Side of the Moon*. Barbara's phone went off simultaneously, playing Nat King Cole's "You Are My Sunshine"—she loved sentimental 1940s music. The two phones produced a rather cacophonous concert but Barbara and Gerry seemed not to mind. They were driving home to their suburban row house after having visited a friend who worked as a scientist at the local hospital's psychiatric unit. The sun had almost completely set when the phones went off, and Gerry handed Barbara a rather peculiar-looking pair of pink sunglasses, which he had produced from her handbag. Before putting on the glasses Barbara winked at Gerry, who smiled back and reached into his coat pocket for his own pair, identical to hers. "Are we really sure we want to do this for Tom?" said Gerry, who looked a bit ridiculous with his pink glasses.

When Barbara had finished her maneuver, overtaking another car, she looked at Gerry and replied, "Of course we'll pull this off. We promised him, and I think it's going to be rather exciting—challenging, but exciting. You look ridiculous, by the way."

"Speak for yourself," Gerry replied.

Tom had asked them to participate in a study he conducted that involved several drastic changes in their daily life. Gerry and Barbara had led so far an almost boringly normal life together—that of many young professionals in the suburbs. But now they had to wear special spectacles at specific times of the day, and they had to expose themselves to as little light as possible once the sun had set and to as much light as possible while the sun was up. Tom's technician had provided them with two phones on which he had installed a special application that produced an alarm whenever they should be keeping away from the light and another one to remind them when to seek light. The users could, of course, choose their own ringtones.

Barbara and Gerry had no difficulty in complying with their prescribed protocol during the first weeks. Thanks to daylight saving time, sunset was around the time they got back home from work and dawn occurred about the time they had to get up anyhow. Although they slept without the curtains drawn, they also used one of those new bedside lamps that simulate dawn—"wouldn't hurt on a rainy day," Tom had said. The dawn simulator started out with almost undetectable light levels, increasing the intensity to its maximum over approximately the same duration as the natural dawn outside. An alarm went off when the light reached maximum intensity, just in case the user hadn't already woken up. Of course, Gerry and Barbara also had their programmed phone reminders. Barbara had banned Gerry's phone from the bedroom because she couldn't tolerate the combination of ringtones in the morning. She had selected a gentle start to the day—Keb Mo's "Every Morning," which didn't harmonize with Santana's "Put Your Lights On." So Carlos had to sing on his own in Gerry's study, except when Gerry was away on a business trip.

The last week of September was unusually beautiful—a true Indian summer. They got up at sunrise and had breakfast on their terrace before going to work. They were surprised to find how early they were able to fall asleep once they retreated into their dark bedroom. Tom's technician had provided them with special bulbs that gave off a warm light, similar to that of an open fireplace. They had installed these light bulbs throughout their house—it was the only type of light they were allowed to use. The technician had also given them adhesive pink acetate sheets that now covered their television and computer screens.

The lamps were essential for the upcoming season of short days and long nights. When they were not at home after sunset, Barbara and Gerry wore their pink glasses to avoid contamination by "unsafe" light. Complying with the protocol in midwinter was quite a challenge, but they somehow survived the dark season. They had slept much longer in the past two months than ever before in their grownup lives. During the long evenings, they did a lot of talking. Often they went to bed as early as eight. During those long nights, they sometimes slept in two parts. After the first, "real" sleep, they would hover for some hours in a state somewhere between sleep and wake before going back to a second round of "real" sleep. At breakfast they would then exchange the strangest tales they had experienced when their brains had gone on fairyland journeys between their two sleep episodes.

When spring finally came, they were filled with a hitherto unknown surge of energy. One morning, looking at the trees outside, Gerry mused that he now knew what it must feel like to shoot leaves. When Tom called and said that it was time to get back to their normal, unnatural life, they seriously contemplated continuing the experiment. In the end they did go back to their normal lives, but they kept their indoor lighting and the acetate sheets over the screens. The only things they didn't miss were those silly-looking glasses.

The American psychiatrist Tom Wehr asked himself what would happen to our sleep behavior if we weren't surrounded by all those modern light sources, which allow diurnal beings like us to see the world around us despite the sun having long set.[1] He persuaded people to participate in his experiment: he asked them to lead a normal life after sunrise and before sunset. But as soon as our astronomical light source had gone, he made them retire into pitch-dark apartments: no lights, no television, not even a refrigerator light that would go on if its door was opened.[2] So what is a sighted person to do when suddenly dumped into complete darkness? The best strategy to avoid too many bruises would be to go to bed, which the subjects usually did. They reported falling asleep after lying awake in the dark for a while. But they hardly ever slept through the entire night, which was longer than twelve hours at the time of year when the experiment was performed. The subjects reported waking up several times during the night, often without gaining full consciousness, thus lingering between sleep and wakefulness. I personally believe that this state must have been the perfect time for the birth of fairy tales and sagas, but I don't remember reading about this in Tom Wehr's papers.[3] What I do remember is that the subjects reported after the experiment was over that whoever thought they knew what "rested" meant had no idea what they were talking about.

During my first two postdoctoral years with Jürgen Aschoff I investigated human annual rhythms. You may think that two years is not enough to scrutinize annual rhythms, and if I had done experimental work you would be absolutely right. But in fact I managed to include more than 5,000 years into my study—or rather more than 60,000 months. I collected monthly rates of vital human statistics from 166 countries, focusing on the annual ups and downs of mortality, suicide, and births.

Aschoff had already analyzed many different statistics and had found that most of them show seasonality (they include exotic statistics, like the number of books taken out of public libraries or crime rates involving assaults). After I had finished my doctorate, I took over this project and worked full time on finding sources, writing to statistical agencies worldwide and compiling a database of monthly human statistics as well as climate data.[4] We wanted to find out whether the rhythms in human statistics correlate with environmental factors or whether they are merely a product of our seasonal social life (workload in agriculture, for example, is clearly seasonal). Since agriculture also depends on environmental factors, it was difficult to distinguish between environmental and social influences. That's why I collected as many years from as many countries as possible. The oldest statistics I found date to 1669, and the most recent to 1981.

Throughout this book you've read that practically everything in our body and in our lives follows a daily cycle. This is more or less also true for our seasonal existence. Seasonality increases the farther humans live from the equator, and it is six months out of phase between the northern and the southern hemispheres. In regions that experience drastic changes in photoperiod (including real winters with temperatures below freezing point and possibly also with a lot of snow), the population's entire life is dominated by the seasons—or at least it was in the preindustrial era. It's therefore not surprising that many statistical aspects of human life show seasonality—some for quite trivial reasons. I have already mentioned that people use libraries differently according to season—borrowing more books in winter than in summer, since reading is an indoor activity during long winter evenings. There are, of course, also more skiing and skating accidents in winter. In Europe in the summer, there are more accidents on the way to school (because more students are on the roadways, riding their bikes to school), and more people drown in swimming accidents or suffer from sunstroke. You already know that we sleep longer in winter than in summer. The difference is not lengthy—only around twenty minutes—but it is very systematic and statistically highly significant.

Other seasonalities in statistics are less trivial: children grow more rapidly in spring than in autumn but put on more weight in autumn than in spring; people eat more carbohydrates and less protein in winter and are more depressed than in summer. "Depressed" is a rather grand word for the seasonality in mood of the general population. Researchers have used questionnaires that are normally used to diagnose depression and its severity in potential patients in the general population at different times of the year and found that depression ratings go up in the autumn and come down in spring. The poll was carried out in response to a syndrome called *seasonal affective disorder* (SAD). SAD patients become severely depressed every autumn and recover during spring. Unlike other forms of depression, in which patients often feel too low to eat a proper meal, SAD patients develop a carbohydrate craving when the days get shorter.[5] In their seasonal moods and eating habits, SAD patients show a pathological extreme of what the normal population goes through in the short days of winter.

Another seasonal rhythm evident in human statistics concerns the rates of suicide. If I asked you to guess in which month the annual rhythm of suicides had its peak, you might guess November or December—that, at least, is what most people think. Given that the seasonality of mood reaches its low point around that time of year, it's a fair assumption. The data, however, show something else. The maximum number of suicides worldwide occurs around the summer solstice.[6] The hypothesis that tries to explain this counterintuitive fact is only indirectly linked to the seasonality of mood. It assumes that when everyone is feeling low, the drive to die by suicide is also lower than at times when everyone else is in excellent spirits. In addition, patients who suffer from bipolar depression and die by suicide do so during their manic phase and rarely during their depressed phase.[7] The actual act of suicide (and not merely the thought about it) takes a level of energy that depressed individuals cannot muster during their most depressed times of year. Thus those individuals who are desperate enough to put an end to their lives do so with higher probability in midsummer, when their energy is highest.

In addition to the seasonal regularities that are evident in the statistics for suicides there is a systematic relationship with latitude. The farther from the equator that humans live, the higher the overall rate of suicide, and the larger the difference in suicide rates between summer and winter. This latitudinal cline supports the notion that the annual rhythm of suicide is associated with photoperiod. The longer and brighter the days in the year, the more energy people have. If they haven't overcome their depression, they are more likely to kill themselves then than at other times during the year.[8]

Photoperiod is one of the major environmental factors correlating with several seasonal statistics. Another player is temperature. General mortality, for example, also shows a seasonal rhythm that has two peaks every year, the major one in the coldest months of winter and a secondary peak in the hottest months of summer.[9] These peaks are identical in the two hemispheres, unlike the seasonal rhythms that show only one peak, which are six months out of phase between the northern and the southern hemisphere. Interestingly enough, the turnaround point between the two hemispheres is not exactly at the equator but approximately five degrees north. The reason for the shift of this "biological" equator to the north is the asymmetrical distribution of landmass across our globe. Landmass is larger in the northern than in the southern hemisphere. The fact that the annual rhythms in vital human statistics reflect this biological equator strengthens the hypothesis that environmental rather than social influences form the basis of these rhythms.

Among the many vital human statistics, monthly birth rates show the most pronounced seasonality that systematically depends on latitude. Many researchers have investigated the phenomenon of seasonal birth rhythms, based on smaller databases and focusing on more limited geographical regions. The authors of these studies have concentrated predominantly on the fact that different numbers of children were *born* at different times of the year; hence they hypothesized that these rhythms correlate with factors associated with the time of birth—for example, with influenza epidemics. When I looked at the regularities in the worldwide database, it soon became apparent that any environmental model made more sense if associated with the time of conception, rather than with the birth date itself.

After I had tested numerous weather and climate factors, a clear picture emerged: what could best explain the influences on the seasonality of births (or rather conceptions) was an environmental model incorporating both photoperiod and temperature. The highest annual increase in conceptions was found to occur around the spring equinox in the statistics from all geographical regions, independent of latitude (except in equatorial regions—I'll come to that later). However, at what point in time the annual conception rhythm reached its actual maximum depended on temperature, and therefore also on latitude. Conception maxima coincide with those times of year when the monthly averages of the daily temperature minima are around 12°C.

The biggest amplitudes in annual conception rhythms were found around thirty degrees of northern latitude, in countries like Jordan, Israel, Palestine, Lebanon, and several regions in northern Africa. In these countries, the difference between the lowest and the highest monthly birth rates used to be more than 60 percent. A reason for the large amplitude of the birth/conception rhythms in these regions might be that the optimal daily temperatures for conception occur close to the spring equinox, so that the two influences—photoperiod and temperature—act simultaneously, resulting in a conception peak at around the 22nd of March.[10] With increasing latitude, the days with optimal temperatures occur progressively later in the year and so do the peaks of conceptions. The largest increase of conceptions still occurs at around the spring equinox, but the actual maximum is reached later.[11]

Whether the seasonality of human reproduction depends on environmental factors or is just a reflection of agricultural workload is still being disputed. One argument of the sociocentric advocates is the distinct difference in the shape of the conception rhythms between Central Europe and the Americas. Whereas the former shows only one pronounced peak (with a smaller one around Christmas), the latter has two more-or-less equally large maxima per year (a *bimodal* rhythm). One fact is often overlooked in this argument, however: similar bimodal rhythms are found in Eastern Europe. The differences in conception rhythms between Central and Eastern Europe (and the Americas) is far more likely to be associated with the respective climates. Central Europe is under the influence of the Gulf Stream, producing a rather temperate climate, whereas the Americas and Eastern Europe have a typical continental climate, with both hot summers and cold winters. Since these summer and winter temperatures are well outside the optimal range for conception, the rate of conception would decrease twice a year, resulting in a bimodal rhythm.

Irrespective of social or environmental factors influencing human reproduction, conception rhythms are apparently not caused by planning, since the seasonality in illegitimate births is in all countries we tested even stronger than those of legitimate births.[12] The fact that conception rhythms in towns are weaker than in rural regions doesn't help to distinguish between social or environmental influences because the seasonality in workload is stronger in the countryside, as are the influences of environmental factors like temperature and light.

So far I have described the systematic yearly cycle of conception rhythms and their geographical distribution only as they existed before industrialization drastically changed human life. Over the past hundred years these rhythms have almost disappeared, and the small remaining conception peaks (mostly no more than 1–2 percent above the annual mean) have changed their phase from spring/summer to autumn/winter. The typical increases with photoperiod and the peaks associated with the optimal temperatures have been replaced by a small peak at around Christmas. This remaining maximum could have purely social reasons—the longer nights and the usual holidays around that time of year lead to more time spent in bed, and thus could slightly increase the chance for a successful conception.

The modern decline in the seasonality of human reproduction can be associated with the onset of industrialization in each respective country—for example, earlier in Germany than in Spain. The next graph illustrates this development. It shows a continuous record of monthly birth rates in Spain over almost the entire last century. Important social events like wars were only able to perturb this immensely regular rhythm as mere blips. After the Second World War, the amplitudes of the rhythm became much smaller. But it was not until the 1960s that the rhythm started to become highly irregular and also shifted its peak from late spring to early winter. Franco had launched a political campaign around that time to massively indus-trialize rural Spain.

Again, these developments cannot distinguish between the social and the environmental explanation. Although industrialization dras-tically decreases the seasonality of the workload, it also means that humans are increasingly shielded from environmental factors. The strongest support for environmental rather than social hypotheses is provided by the observation that the two factors—photoperiod and temperature—have lost their influence in a systematic way. The in-fluence of photoperiod disappeared first. In the early days of in-dustrialization, people stopped working outside and began working inside, which shielded them from daylight. They still were exposed to seasonal temperature fluctuations, however. An environmental model using only temperature as a predictor can fully explain the an-nual conception rhythms of this era. Only much later were people shielded with comparable efficacy from the influence of temperature by both central heating and air-conditioning. It was only then that

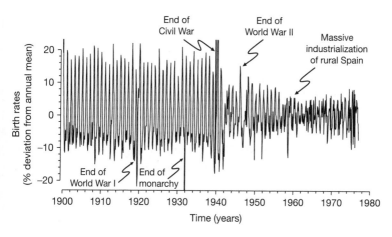

A record of monthly birth rates in Spain over almost the entire last century. Reprinted from T. Roenneberg (2004). The decline in human seasonality. *Journal of Biological Rhythms* 19(3):193–195.

human reproduction almost completely lost its seasonality, except for the aforementioned (social) blip around Christmas.

One could argue that the contraception revolution brought about by the birth control pill coincides with the observed loss of rhythmicity in human reproduction. Yet the introduction of the pill had astonishingly little influence on the seasonality of reproduction—only on its level. This is not too surprising because contra- and proconceptive planning are both part of sexual behavior, which is seasonal, as the annual rhythm in condom sales clearly shows.

We don't yet know the underlying mechanisms for the seasonality in human reproduction. Is there seasonality in both male and female sexual behavior? Is it that sperm production or sperm agility is seasonal, as studies have shown, or that eggs are more or less fertilizable? Or is there seasonality in how successfully a fertilized egg survives the first weeks of pregnancy?[13] The following finding supports this last possibility: although some illnesses like schizophrenia are only diagnosed in early adulthood, the births of these patients are seasonal—showing an even stronger amplitude than those in the general population. Many of these illnesses have a genetic basis, which could already have an influence on the development of these patients as embryos. If embryonic development were seasonal, then these patients would have a higher chance of being born at certain times of the year than at others, which could explain the annual rhythm in their birth rates.

It is quite remarkable how many aspects of our lives are influenced by industrialization and its consequences, isolating us increasingly from natural signals. To allow organisms to accommodate and anticipate the natural changes across day and night but also across seasons, evolution has developed internal clocks. One might argue that an industrialized human doesn't need these clocks anymore since we live in a society that is not only 24/7 but to some degree also 24/7/365. Yet we shouldn't forget that life in any form is constantly under attack. Where the attackers of our ancestors included big enemies (like saber-toothed tigers, wolves, or bears), our modern selves are still under attack, only by much smaller but just as dangerous enemies like bacteria, fungi, or viruses—and these enemies are still seasonal. An appropriate timing system in our body would therefore still be useful—at least for our immune system. But our internal clocks are only effective if we provide them with sufficient information: with sufficient differences in daily light and darkness, or with sufficient contact to changing photoperiods and temperatures.

Tom's experiment made Barbara and Gerry live strictly by photoperiod, and neither of them developed a cold that winter.[14] You probably guessed the rationale behind the pink glasses, the filter sheets, and the special light bulbs. After dusk and before dawn, they aimed to shield Gerry and Barbara from that part of the light spectrum that reaches our timing system most effectively (the blue parts of the light spectrum).

23

The conference dinner had long been over. About thirty neurosurgeons had gone to a jazz bar in the old part of town to celebrate a successful meeting. The core of the party consisted of three neurosurgeons—among the best in the world—who had known each other for decades. Their frequent joint appearances had earned them the nickname "the three-pack." Many of those who attended these conferences regularly had learned over the years to follow wherever the three-pack led—especially when Drs. Fergusson, Skinter, and Lafayette went to a jazz bar.

A quartet (bass, saxophone, piano, and drums) was playing swing on a small podium in a corner of the room opposite the long bar. The group of surgeons had moved several of the small tables together between bar and podium. By now they were the only patrons left in the bar. When the musicians stopped for a break, Fergusson made eye contact with Skinter and Lafayette. "What do you think? Shall we go for it?" Lafayette's face lit up with almost childish anticipation, but Skinter looked, as always on these occasions, far too tired to respond adequately. Fergusson ignored Skinter's familiar lack of enthusiasm and got up to chat with the musicians, who were having a drink at the bar. When he returned, he nodded to Lafayette and Skinter and the three went up onto the podium. They almost had to carry Skinter to the piano, where they dropped him onto the stool and placed a triple espresso on a small table beside him. Fergusson climbed behind the drums, and Lafayette picked up the bass.

"A one, a two, a one, two, three, four," and off they went on a journey of the coolest swing, blues, and boogie-woogie ever heard outside of a concert hall. The triple espresso seemed to kick in, carrying Skinter along despite it being so much past his usual bedtime. He was one of the best jazz pianists around, if one of the least known— few people outside of the neurosurgical world had ever heard of him or had ever heard him play. But within that circle he was famous for both his innovations in neurosurgery and his heavenly command of the piano. If it had been several hours earlier, he definitely would have brought the house down and would have made sure that the bar's regular quartet, though excellent themselves, would have retired for the night. He could, without doubt, have made it to the top of the crop in the jazz world. But one just cannot live two full lives within one lifetime.

So far I have discussed the science and history of clock research. Although the stories at the beginnings of chapters wrap the science in fiction, the content of every chapter is based on peer-reviewed, published fact.[1] In the remaining two chapters, I take the liberty of engaging in more speculative thoughts about body clocks. In this chapter I will discuss how chronotype and the individual need for sleep could influence people's careers, or even their behaviors and personalities.

This chapter's case story was inspired by a colleague of mine who is both a gifted neurosurgeon and an extremely talented jazz pianist. He could have chosen either profession and would have been very successful in both—that is, if he hadn't been such an extreme early chronotype. He would not have survived the long concert evenings, having to give his best performance way past his peak abilities and preferred bedtime. He has definitely made the right choice in his career because he has no difficulty whatsoever in giving his best performance in the operating theater at 6 or 7 A.M. The early working hours of doctors suggest that this profession is submitted to a selection pressure favoring early types. In addition, doctors often have to cope with sleep deprivation, especially at the beginning of their careers, which potentially has consequences for patients.[2]

A few years ago, Harvard sleep researchers investigated the medical errors made by young interns. They compared the error rates in two types of work schedules: the traditional, extended schedule and a so-called intervention schedule. During their study period, the Harvard group investigated a total of 2,203 patient-days involving 634 admissions and found that interns made about 36 percent more serious medical errors on the traditional schedule. Disturbingly, more than half of these were not intercepted before reaching the patient.[3] The results of this study are perhaps to be expected. It's not surprising that someone who works for up to seventy-two hours without noteworthy rest makes more mistakes, but what seems like common sense is worthless without a quantitative basis provided by a properly designed study.

After the results of that study were published, decision makers reluctantly started to implement slightly better work schedules for interns. Before the Harvard sleep researchers published their data, no one was able to change the system, even though everyone believed that grueling schedules must produce more medical errors. Why is this? First of all, we are dealing with a very special portion of society. Doctors have gone through a long university education, their responsibility concerns the life or death of a patient, and they remain among the highest-ranked professionals on our ladder of social recognition. It seems that this elite system actively exerts a selection pressure that ensures the survival of only the fittest. Besides, those who make the decision about a young doctor's working hours went through exactly the same extreme work schedules when they were interns themselves and don't see why this tradition should be changed.

Physicians are a prime example of professional selection involving the capacity for short sleep and top performance in the early hours.[4] Teachers fall into this category as well. To discipline, reach, and teach between twenty and forty children or teenagers at 8 A.M. requires teachers to be in full command. Being a teacher is one of the few professional situations where individuals remain in the same environment (except for the interruption of going to university) from childhood to retirement. This suggests that they felt comfortable with this environment before they made their career decision. It seems unlikely, therefore, that they are particularly late chronotypes, or even were late chronotypes as teenagers, and thus may belong to the early chronotypes in our population.[5] Note that all of this is purely hypothetical. We need to conduct epidemiological studies into professional selection, and doctors and teachers would certainly be a good starting point.

The selection for early types and short sleepers may be a general prerequisite for the top professions in our society, for managers and decision makers who have to give their best from early in the morning until late at night. Can you imagine a manager or a politician who needs an average amount of sleep and is an extreme late type? Such an individual might never make an appointment before 11 A.M. A late type and long sleeper would also have a higher statistical chance of falling ill, since lack of sleep challenges the immune system.

A recent study recorded the sleep duration and sleep quality of about 150 healthy subjects for two weeks before the participants were quarantined in a clinic and given nasal drops containing a rhinovirus. The researchers then looked for the development of clinical symptoms of a cold. Participants who slept seven hours or less were almost three times more likely to develop a cold than those who slept eight hours or more. The influence of sleep quality was even stronger than that of sleep duration. Individuals reporting their sleep quality to be less than 93 percent of the possible optimum had a five-and-a-half-fold chance of catching the cold compared with individuals who rated their sleep quality to be 98 percent of the optimum (or higher). Of course the researchers made sure that this result was not caused, for example, by the fact that subjects slept badly because they already were fighting off an illness during the baseline period.[6] Thus, poorer sleep quality and shorter sleep duration in the weeks preceding exposure to a rhinovirus were associated with lower resistance to illness. This result strengthens the notion that decision makers are more likely to be early types and short sleepers, simply by selection. Their sleep behavior efficiently protects them from falling ill, thereby making them overall more successful compared with late types and long sleepers, who would regularly suffer from social jet lag if they had to comply with a decision maker's daily schedule.

On several occasions in this book, I have described the difficulties people have in taking seriously the body clock and its biological impact on our lives. These include the elderly professor at the medical school, who thought the body clock was surely only important for sensitive people; or those who insisted that one could readily adapt to certain working hours if one only showed some discipline; or the politicians and teachers who were convinced that young people could not get to sleep early enough to function properly in the morning only because they watched too much television or hung out in discos. Are these critics mostly early types and short sleepers? If so, it is not surprising that the insights of chronobiology are so sluggishly applied in everyday life—the decision makers have no experience with the social jet lag others suffer from, and therefore see no need to change the system. This may be similar to the reason why so many railway stations or public buildings are so poorly adapted for individuals who have difficulty walking—most decision makers can walk perfectly well.[7]

Napoleon is supposed to have said, "Six hours' sleep for a man, seven for a woman, and eight for a fool." Napoleon shared the ability to get by on little sleep with many other leaders—even dictators like Hitler or Stalin. The inventor, scientist, and businessman Thomas Edison claimed that sleep was "a waste of time."[8] But then he apparently also did take frequent naps during the day, an ability he shares with many managers and politicians. Some of them have a remarkable capacity for naps (like cats and dogs)—they can fall asleep within seconds whenever they find a free moment; for example, when sitting in the back of a limousine that takes them from one commitment to the next. It is worth investigating whether the ability for short sleep may be associated with a personality showing tendencies to mania. But not every influential personality regards sleep as a waste of time. Einstein apparently required at least ten hours of sleep for optimal performance.

I once witnessed a good example of sleep phobia coupled with mania at a sleep conference. A physicist had somehow convinced the organizers to offer a talk about the "myth" of having to sleep eight hours. In his lecture he tried to convince his audience that we have become mollycoddled and somehow brainwashed into thinking that we need on average seven to eight hours of sleep. He claimed that we could, with the help of a short training program, reduce our sleep-need without deleterious consequences to about three hours. In doing so he ignored all knowledge about the biology of sleep and its genetic basis. Any biological quantitative trait has a distribution in a population. In denying that sleep duration has a genetic basis, he might just as well have claimed that sleep is not a biological trait in the first place. A paper published in the journal *Science* in August 2009 clearly shows that some short sleepers have a mutation in one of the clock genes.[9] This proves that there is indeed a genetic basis for individual sleep duration. The sleepless physicist was so wrong! The next morning at breakfast, he provided living proof of the falseness of his claims. I witnessed an encounter between him and a waitress: "I can definitely tell caffeine-free coffee from the real thing, and what comes out of this pink thermos is pure decaf, although the label claims it's coffee!" He was beside himself. Which brings us to the relationship between the body clock and personality or behavior.

Among the first functions to go as a consequence of sleep deprivation are the social skills—an agreeable personality obviously depends on good sleep within our personal circadian window. Most of us have experienced this with young children. When they haven't slept, their behavior is remarkably obnoxious—so bad in fact that inexperienced parents, relatives, and onlookers might entertain the notion of taking them to a psychiatric unit. Once they have slept, they revert to their angelic selves once again. This is exemplified by a fascinating medical case involving a severe but fortunately rare genetic disorder. Smith-Magenis syndrome is responsible for a multitude of symptoms: mental retardation, malformation, and congenital anomalies, but also severe behavioral pathologies.[10] Smith-Magenis children are extremely difficult to handle, have frequent temper tantrums, are extremely tired during the day, and have difficulties sleeping at night. Scientists recently discovered that some Smith-Magenis patients, especially those with the behavioral symptoms described above, have an inverted melatonin rhythm. Normally this hormone is produced at night, but in these patients, it comes up during the day and is absent at night. French pediatricians treated Smith-Magenis children with a beta-blocker in the morning and supplemented melatonin in the evening, thereby reverting the abnormal melatonin rhythm back to its normal phase.[11] The result of this treatment was quite spectacular: many of the behavioral problems disappeared, and the children were able to sleep better at night. What was originally thought to be a direct symptom of the genetic deletion turned out to be more of an indirect symptom of children who were forced to sleep and be active at the opposite ends of the twenty-four-hour day. Think of how other children would behave if we wanted them to be active all night and forced them to sleep only during the day.

So chronotype and the individual need for sleep apparently have an effect on our lives and our choice of career; in addition, they seem to be associated with personality and behavior. But which is cause and which is effect? Does personality lead to a certain chronotype or sleep type, or does chronotype or sleep type lead to a certain personality? Several studies have investigated possible associations between time-of-day preferences and personality traits such as extroversion, agreeableness, conscientiousness, neuroticism, or openness.[12] Some of these studies found that morning people are generally more agreeable and conscientious while evening people appear to be more neurotic, extroverted, and innovative. Association studies have also looked at sleep duration and found that longer sleep makes individuals more agreeable.

If we were to take the results of these association studies at face value, a clear picture seems to emerge. Late types are exciting individuals, albeit a bit grumpy, extroverted, and neurotic, especially when young, and on the whole they belong to the more innovative members of society. In contrast, early types are very agreeable creatures, reliable and conscientious, but, alas, a bit boring.[13]

Categorizing humans into boxes based on associations is an old but rather dubious pastime of scientists; it goes back to Hippocrates, who classified human temperaments into four categories based on different body fluids.[14] Another typologist was Ernst Kretschmer, who used the physical appearance of the body to classify human temperaments.[15] The trouble with all these classifications is that they are *a priori* based on existing prejudices. Hans Eysenck was the first scientist to try to eliminate the influence of prejudices by using a more scientific approach based on statistics.[16]

People with certain daily preferences (*not* equivalent to a chronotype determined by the actual times of sleep and activity) are loosely associated with certain personalities, but such associations may be based on a circular process. People who have a strong urge to comply with social rules, for example, would be biased when filling out a questionnaire probing their daily preferences. They might hesitate to admit that they are late types (under the pressure of public opinion) and would thus artificially create an association between personality and morningness-eveningness type. Do late types appear more innovative than early types because they were, for example, more challenged in school than early types, and always had to invent clever strategies that helped them perform despite not being on top of things in the morning? Do early types choose standard, less innovative careers because they had better grades at school? The fact that early types are more agreeable than late types could merely reflect their low level of social jet lag. As chronobiologists, we would at least demand that personality questionnaires be filled out at different times because the answers would surely depend on time of day.

There are far too many different proffered explanations for the associations found between daily preferences and personality; this is always a sign for a lack of understanding about what the results mean. More, and more rigorous, research is required, using approaches that can tease apart the many potential influences and causalities. Any such investigation should start with an assessment of chronotype based on actual sleep and activity or even more objective measures, such as daily melatonin profiles, instead of the extremely subjective morningness-eveningness preferences.[17] It should also take into account at what time of day the questionnaire was filled out in relation to the subject's chronotype. Finally, such a study should note a subject's profession, and whether or not the career was chosen fairly freely. My prediction is that if associations between chronotype and personality exist, they may be quite indirect, depending on how different chronotypes fit into the temporal demands of society. Imagine we had the following combination of four individuals, who all shared the fictional Dr. Skinter's talents, but who differed in their chronotype and in the profession they chose (or were forced to choose): the first might be an early type who has become a neurosurgeon; the second, an early type who has become a jazz musician; the third, a late type who has become a neurosurgeon; and the fourth, a late type who has become a jazz musician. How would these four fill out the questions of the Big Five Inventory? The analysis of such a combination of cases would allow us to tease apart whether personality is directly linked to chronotype or whether their association is an indirect result of how different individuals fit into the temporal demands of society.

24

The three older boys always thought that their slightly chubby, red-haired, freckled-face little brother Leon was a bit peculiar because he seemed more interested in strange self-experiments than in kicking a ball with them in the garden. Leon's most recent project was to find out how the brain works. It had started last Monday when Professor Mallet, the father of a classmate who was a scientist at the local university, had talked to their biology class about how the brain functions. From that day on, Leon had been looking at the world through different eyes. He began to think of many projects that might help him to see how his brain sees the world. The fact that he could use his brain to think about himself intrigued him.

He started to investigate optical illusions that fooled the brain into making mistakes. Lines on paper could be absolutely parallel—Leon established this with a ruler—but if other lines were drawn across them at an angle, they stopped looking parallel. Professor Mallet had told them that if only one single step of a flight of stairs were of a different height, people tended to trip. The brain registered the height of the first few steps and then assumed that the heights of the others would all be identical. Leon entertained the idea of secretly changing the height of one of the steps in their staircase at home by sliding a thin board under the carpet; but since not only his brothers but also his parents frequently ran down those stairs, he quickly abandoned the idea.

He took to closing his eyes whenever he went into a new room, trying to "see with his inner eye" all of its facets. Where was the door? Where were the windows, the table, or any other object he could re-member? One day he designed an obstacle course in the basement playroom that was to be experienced in the dark. This time his brothers were interested enough to join in his antics, but with the lights switched off, they stumbled into things. And so for once Leon beat them in every single trial.

On the weekend he visited his aunt. When he woke up in the morning, he was completely confused—the sort of confusion you may also have experienced. He went to look out the window but was confronted with a wardrobe. He looked at the door, but there was only a solid wall with a rather silly picture of a flowerpot. Where was he? What was going on? After what seemed like a small eternity, he suddenly remembered that he wasn't in his own room at home but in his Auntie Gwendolyn's guest room. Immediately everything fell into place. Another project had fallen into his hands! What tricks had his brain been playing on him? Had it expected to wake up in his bed-room at home? Why had it taken so long to recognize its mistake? Now that his confusion had cleared, he closed his eyes and was able to "see with his inner eye" both his own bedroom and Aunt Gwendo-lyn's guest room.

Back home, Leon's thoughts became more ambitious. He re-flected on something else Professor Mallet had told them: the brain had quite a stubborn character of its own and wasn't just prepared to passively see, hear, feel, or smell. It could never look at something without some kind of prejudice. He realized that the world around him was somehow etched into his brain. The brain must be con-stantly building an internal model of its surroundings—similar to the paper models of ships or houses he had built. This made excel-lent sense, thought Leon, because the brain could then focus only on the things that changed. This gave his brain more time to do other things, like thinking about itself.

Leon eventually got bored with investigating how his brain dealt with space and started to explore how it dealt with time. His grand-father had given him a really cool wristwatch for Christmas—his first—and Leon wore it day and night. Previously, Leon had always seemed to know the approximate time of day. Now he caught himself constantly consulting his watch. He might calculate, for example, that there would still be enough time to finish his homework before leaving for his violin lesson, but then he would immediately forget, and he would check the time on his wrist every five minutes. He decided to leave the watch on his desk and try to consciously teach his brain to estimate time as accurately as possible. He was particularly proud to find that he could even get the time right during the night but then realized that he always awoke at the same time—between 2:30 and 3:00 A.M. He tested whether he could make himself wake up an hour earlier and was surprised that he succeeded after only a few attempts. But there was something puzzling about how his brain figured out what time it was in the night. He never had a good sense of how long he had been sleeping, so obviously his brain didn't judge the hour by how much time had passed. It just knew! He wondered if it created a similar mental model of the twenty-four-hour day as it obviously did for space.

This book is about clocks. Not about those you can buy, wear, or hang on a wall, but about the clock that ticks away in your body. These were the first two sentences you read in this book, if you are the type who actually reads introductions (I must confess that in most books I don't). You might therefore have wondered why I now, in the last chapter, tell a story that predominantly deals with space. But as you will see in this chapter, space and time have a lot in common (well beyond Einstein's space-time continuum). The reason for calling our body's daily timing system a clock is obvious: its function appears to be measuring time of day. The word *clock* has very much shaped the way chronobiologists investigate this phenomenon (all our thoughts are strongly shaped by the language we use), but it is a severe reduction of the many services that our internal timing system provides.

By analogy, we could call the capacity of the brain to create a virtual copy of the space around us—the one Leon was so fascinated by—a compass. Our brain's virtual space machine can certainly be used as a compass because it helps us orient ourselves in space, even if someone switches off the lights.[1] But there is much more to this virtual space machine than orientation: the entire structure of our environmental space is kept in working memory. Leon and his brothers raced through an obstacle course in a dark room after having memorized all its structural details. This ability obviously didn't incorporate only knowledge of the direction of the finish line (such as a compass would provide) but also the exact locations, distances, sizes, and shapes of the obstacles.

The body clock is a similar representation of a temporal space, a *Zeitraum*.[2] In this case it represents the *Zeitraum* day. In several functions, the body clock is used as a compass as well as a clock. In a wonderful series of experiments, the late Eberhard Gwinner was able to demonstrate how migratory birds use their body clocks for navigation during their long voyages.[3] He kept birds in big aviaries exposed to the natural environment. During their migratory season, he placed birds at regular intervals in funnel-shaped cages with walls of paper and an inkpad at the narrow bottom. After an hour or so, he had a perfect record of the direction in which the bird was predominantly active. All he had to do was to analyze the density of their black footprints on the paper. The results were stunning. While the free members of the species flew from Bavaria to the South of Spain (on their way to their Central African breeding grounds), the experimental birds that were kept behind in the Andechs aviaries hopped predominantly in a southwesterly direction. Once their free conspecifics had crossed the Strait of Gibraltar, the experimental birds also changed their direction toward southeast—as though they were now flying toward Central Africa. We don't have to presume some esoteric communication between the actual travelers in Spain and the virtual travelers in Andechs because the birds have an internal program that tells them to fly approximately x days in one direction and then y days in the other. The virtual travelers were able to copy the behavior of the actual travelers only if the top of the funnel-shaped cage gave them a free view of the sky. So they obviously oriented themselves by using the sun (or the stars) as a compass. Of course, the sun and the stars don't stay in one place, because our globe turns. To be able to meaningfully translate the position of a celestial object into a constant direction, the birds had to know the time, and Gwinner was able to prove that they use their body clock for this task. He took some birds out of the natural aviary and kept them for a week in a room where he resynchronized their body clock to an artificial light–dark cycle that was out of phase with the local Andechs day–night cycle. When he then placed the birds into the funnel-shaped recording device with free access to the local sky, they hopped in the wrong direction (although it would have been the right direction according to the resynchronized body clock).[4]

The ability of some people to program themselves to wake up at a certain exact time is another example where the body clock is used in its literal function. Although this ability is not part of the body clock itself, it uses its "knowledge" about time of day. (Leon used this ability to reset the timing of his regular nocturnal awakenings.)

But the biological timing system can do much more than simply tell the time of day: it internally represents the *Zeitraum* day. It forms a network of information between biochemical processes at the cellular level and the behavior of the organism. Thereby it creates an internal temporal structure that coordinates all bodily functions within the *Zeitraum*. This temporal program anticipates the regular environmental changes and prepares the organism to do the right things at the right times of day. The different stages of the day and their different challenges (light and darkness, warmer and colder temperatures, as well as time of day–dependent availabilities of food sources, presence of enemies, and so on) are comparable to the obstacles in the races organized by Leon.[5]

Evolution itself is an even more compelling case for the hypothesis that the body clock is not simply a time-of-day teller. As we have already discussed, the rules of evolution are based on (almost) random mutations, which are then challenged by many different selection pressures. An important selection pressure is that of competition for resources. These need not necessarily always be food; sufficient space and opportunity to reproduce and raise offspring are as much resources in the selection process. This is why all living creatures constantly search for new niches offering them potential advantages during their evolution. Conquering a niche often involves genetic changes across a species that allow better adaption to the new environment.[6] Pioneers occupying a new niche have an enormous advantage until the competition arrives. We commonly think of these niches as spatial structures. When the first animals left the oceans and survived on land, many adaptive changes had taken place: for example, legs had been selected for over fins, and lungs over gills; the body had to cope with much more weight (no floating around anymore); measures had to develop to prevent desiccation as well as to find and store water.

Humans are primates, and primates are mammals. The first mammals appeared on earth between 200 and 250 million years ago—a very short time in evolutionary units. To put that figure in perspective, the first primitive unicellular organisms appeared roughly 4,500 million years ago; the first cell with a proper nucleus appeared about 1,500 million years ago;[7] and the first animals with bones inside their body appeared on land only 380 million years ago.[8] These were the ancestors of reptiles, which started to conquer every livable niche outside of the oceans. But after that, where could one turn when it got too crowded on land? There was one spatial niche left that hadn't been conquered yet by larger animals: the airspace above, so far colonized only by insects. The first flying reptiles pioneered this niche (the ancestors of the feathered creatures we know today as birds). But there was yet another niche waiting to be conquered.

Life on land also meant that animals often had to cope with the large temperature changes between cold nights and hot days.[9] Everything in a living creature depends on biochemical reactions, and most of them happen faster the warmer it is. This means that reptiles can catch prey and run away from enemies more easily during the daytime, because the biochemical reactions necessary for movement function more efficiently. No wonder that practically all reptiles are day-active. Being day-active means that all aspects of life adapt to the sunlit day, from basic metabolism to the senses and behavior.

You may have asked yourself what niche could have been left after ocean, lakes, land, soil, and airspace had been occupied. But now that I have mentioned the alternatives of night- and day-activity, you may guess where this chapter is going. This particular hypothesis about how mammals evolved presumes a change not between two spatial niches but between two niches within the *Zeitraum* day. While birds are the end product of an evolutionary line that conquered the airspace, the ancestors of mammals conquered the night. Our mammalian ancestors went through a nocturnal bottleneck. Developing the ability to fly was a prerequisite to conquering the airspace, thereby happily feeding on insects and getting away from all those ground-bound reptilian enemies. Perfecting temperature regulation (becoming truly warm-blooded) was a prerequisite to conquering the cold nights. During the day our mammalian ancestors could hide in a dark and cool burrow and during the cold nights they could roam around, dodging dangerous but now potentially sluggish reptiles.[10] Mammals could keep their body temperature at around 37°C even at night while most reptiles were highly dependent on the ambient temperature—they were simply not competition.

Changing from one spatial niche to another would not be possible without an excellent internal representation of the environment. Analogously, switching between niches within the *Zeitraum* day would not have been possible without a comparable internal representation—provided by the body clock. It seems that during evolution changes between spatial niches and those within the *Zeitraum* day never occurred simultaneously—confirming their equivalent importance for the organism (the long evolutionary adaptation process would not allow the conquering of two independent niches simultaneously). Changing a niche is a long-term prospect, but it is not a one-way street. Many terrestrial animals have gone back into the water and many birds have forgotten how to fly. Switching back to an original niche also happens in the temporal domain. All our ancestors must have been night-active (most mammals still are), but we and some other mammals have reconquered the day. Once the day-active birds had firmly established their dominance in the air, some of them switched to night activity. The lark is a good example of the former and the owl is a good example of the latter. You see, this book is about larks and owls from beginning to end!

Our internal clock is not only important in making our body do the right things at the right times; it has also been a crucial player in our very own mammalian evolution. In this chapter you have seen the great advantages of flexibility in occupying niches on an evolutionary scale. The wide variety of chronotypes in modern populations potentially provides similar flexibility. Our species has conquered almost every spatial niche on earth. We haven't adapted to all these niches as organisms do in evolution, however. If we couldn't carry with us all we need as day-active, land-living mammals, we would not survive. We need torches in the dark, we carry oxygen under water or when the air is too thin, and we heat our homes in winter, to name just a few of our adaptive insufficiencies. Our flexibility within the *Zeitraum* day is, however, genuine—let's make the most of the chronotype differences within our species.

Ann and Jacob would perform so much better if we appreciated that most teenagers have to start school at around their internal midnight. If we could sleep within the appropriate circadian window (like the young man in the bunker, or like Barbara and Gerry—but unlike Sergeant Stein or Timothy) we would be less tired and more cheerful during the day; we would perform better; and we would be healthier. Our work schedules have to acknowledge that most of us are no longer farmers. More flexibility would be advantageous for gathering the proverbial mushrooms or for hunting at night like Urf. They would also allow people like Dr. Skinter a real choice in their professional careers. We would even become more tolerant in our moral judgments about the faster and slower hamsters among us, and spouses, like Louise and Bruno, might be more forgiving of their respective differences.

Internal timing is genetic, as Sarah and her family would testify. In addition, Harriet would sadly agree that our body clock doesn't simply respond to social cues but uses its own response characteristics for synchronizing to day and night, light and darkness on earth (and not on other planets). That's why Germany's former socialists go to work earlier than their capitalist compatriots. We have to appreciate that industrialization means working inside and that lack of light is problematic not only for our body clock but also for many other aspects of our well-being—think of Hanna, Sophie, Frederic, and Joseph. Architects have to find ways to allow more bright daylight into buildings (without increasing our carbon footprint).

As we are likely to increasingly abuse our body clock with jet travel and shift work, like Oscar and Jerry or Marco and Maria, basic and applied research has to offer solutions for minimizing the detrimental consequences. We need new individualized shift-work schedules. We have to discuss the pros and cons of messing with our body clock as we do with daylight saving time (although our motives should be more constructive than Edgar's). Internal timing is of ecological importance—from single-cell algae to man—and was instrumental in evolution. It also controls all of our bodily functions and therefore should play a crucial role in medical diagnosis and treatments. Although our curious astronomer acknowledged the wide consequences of his discovery, I doubt that he had the slightest idea of how colossal they would turn out to be.

Notes

1. Worlds Apart

1. A moral from James Thurber, *Fables for Our Time.*
2. By definition, the term *sex* is used to distinguish the difference between a man and a woman based on inherited genes; the term *gender* is used to indicate differences due to learning, social context, or culture.
3. Age at the onset of puberty varies among different regions and also depends on diet and cultural setting. Nowadays, puberty could be defined biochemically by measuring certain hormones. Practically, however, more evident signs are used. In girls it is defined as the first menstruation (menarche), in boys by the growth of the testes and the penis. The end of puberty is defined in both young women and men when the arms, legs, and feet stop growing (not necessarily when full body height is reached).
4. From the Greek word *chronos.*
5. Long before scientists set out to investigate different chronotypes in humans, people have noted that different bird species start to sing at very specific times. In Central Europe, for example, the redstart is one of the first to sing before the sun rises, followed in incredibly precise intervals of approximately five minutes by the robin, the blackbird, the wren, the cuckoo, the great tit, the chiffchaff, and so on. Unlike the question we pursue in this book, the individual differences *within* a species, this "bird clock" shows the differences in chronotype *across* species.
6. The Munich ChronoType Questionnaire (MCTQ). At the time this book was written, our database contained about 120,000 entries. Initially, we

used paper questionnaires, but now we administer it over the internet. You can answer this questionnaire at the following address: www.the Wep.org (choose "Chronotype Study" once you are past the introductory page). If you enter your email address you will receive a .pdf file describing your chronotype in relation to others. It's best to go to the website before you read too much of this book because naive subjects answer questions more reliably.

7. Originally the science of epidemics was a branch of medicine that dealt with the study of the causes and distribution of disease in populations. Epidemiology has been extended to include the study of normal conditions of humans, such as sleep, eating behavior, and so on. This extension is a logical consequence of the insight that a disease is, in most cases, an extreme form of a normally distributed human quality.

8. To establish the distribution of a quality or a trait, a researcher decides on categories or "bins" with defined limits. In the case at hand, the distribution is based on bins of thirty minutes. Such a bin collects, for example, all midsleep times between 1:30 and 2:00 A.M., excluding 2:00 itself because it already forms the lowest limit of the next bin. The counts in each bin are then graphed, commonly as vertical bars. To compare different distributions that relate to populations of different sizes, one converts the actual counts in each bin to the respective percentage of the entire number of people investigated.

9. Distributions that are completely symmetrical are called *normal* distributions or *bell-shaped* distributions (since they look like the cut through a bell); they are also sometimes called *Gaussian* distributions after the mathematician Carl Friedrich Gauss, who introduced them. If you record every day how long it takes you to get to work, you can calculate the average time of your commute. Of course the times will be slightly different every day. If the chances of being shorter or longer than average were not biased in any way, you would get a normal distribution with a perfectly symmetrical bell shape. You would, of course, always like to get to your workplace in the shortest possible time, and the forces acting against that are not unbiased; they will rather actively slow you down than actively speed you up. If you are slowed down, the hold-ups can be quite substantial in some cases. If the bias against reaching the mean of

a distribution is more toward one side than the other, it will produce a slightly unbalanced distribution, similar to the one in our example. The chances of being a later chronotype than the population's average are slightly higher than those of being earlier than the average.

2. Of Early Birds and Long Sleepers

1. To my mind, the brilliance of Baba Shah's witty statement lies in ridiculing all attempts to move early risers closer to God than other chronotypes and in recognizing that different species occupy different temporal niches within the twenty-four-hour day.
2. Some people think that this proverb refers to golden teeth as a sign of wealth, but in this case, "mouth" merely stands for an opening, equivalent to an open hand.
3. Noon stays constant throughout the year except for some wiggling that amounts to no longer than a minute.
4. For example, see Ian R. Bartky, *One Time Fits All* (Palo Alto: Stanford University Press, 2007).
5. The period length of the lunar cycle is also caused by interactions of sun, earth, and moon.
6. Seasonal changes exist even at the equator (for example, in the amount of rainfall), although day length stays constant at twelve hours throughout the year.
7. This is especially true in German: "Frühaufsteher" and "Langschläfer," early risers and long sleepers.

3. Counting Sheep

1. Modafinil is one of the drugs that make people feel quite normal in spite of not having slept enough. Caution: we still know too little about sleep, so we cannot predict the long-term consequences of suppressing the effects of sleep deprivation. The state we are in when having slept too little is, like pain, part of a warning system that we most probably should not ignore.
2. An oscillator is anything that can produce a rhythm: mechanical oscilla-

tors such as a pendulum or a swing; electrical oscillators such as door-bells; chemical oscillators that produce rhythmic transitions from one chemical state to another and back (often visualized by changing colors). The biological clock is also an oscillator.

3. S. Daan, D. G. Beersma, and A. A. Borbely (1984). Timing of human sleep: Recovery process gated by a circadian pacemaker. *American Journal of Physiology* 246:R161–183.

4. Sleep researchers call this the "wake maintenance zone."

4. A Curious Astronomer

1. As Serge Daan recently pointed out in a lecture on the history of the field, there is no evidence that de Mairan used a desk or cupboard to perform his experiments. This notion probably goes back to one of the first textbooks on biological clocks, written by M. C. Moore-Ede, F. M. Sulzman, and C. A. Fuller: *The Clocks That Time Us: Physiology of the Circadian Timing System* (Cambridge, MA: Harvard University Press, 1982). In this book, the story of de Mairan was illustrated by a drawing showing a mimosa plant placed in a desk.

5. The Lost Days

1. Some of these bodily rhythms included: the alternation between sleep and wakefulness and between rest and activity; the ups and downs of body temperature, hormones, electrolytes, and cognitive performance; and how fast or slow subjective time passed.

2. A Latin-derived construction for "about one day."

3. Notably also the acronym for rapid eye movement, which is the term for the stage in our sleep when we often have dreams, move our eyes, and twitch with our muscles as if we were awake.

4. For example, subjects are shown a light for a certain time (or hear a tone) and have to say how long it lasted; in other tests they have to press a button and reproduce the length of the signal they just saw (or heard), or they are told to produce a given time by pressing a button. In the bunker experiments, subjects were merely asked to press a button when

they thought an hour had passed and press the button again when they thought a minute had passed.

5. The rules of the theory of evolution are simple. An organism's genetic code, its deoxyribonucleic acid, or DNA, holds the information about every biochemical "tool" that cells need for living. Every offspring (of either sexual or asexual reproduction) can accumulate random changes (mutations) in its genetic code. The number of copies of a given genetic code within a population depends on how many offspring inherit this code over a long line of generations. In this "game," individuals compete with other members of their species. Those that pass on more copies to the next generation than others win the game. This competition is under constant pressure: from members of its own species; from other organisms (both enemies and food); and finally from the physical environment, such as heat, drought, or anything else living beings have to cope with. If a random change in the genetic code helps an individual to better cope with these pressures, its chances to proliferate its genetic code become greater; if the change decreases its coping abilities, its chances of proliferation decrease, and that particular genetic code will be eventually outnumbered by others.

6. The Periodic Shift Worker

1. Note that the example of the lonely bunker worker is not the whole truth. The principal investigator, who worked together with Aschoff on the bunker experiments in Andechs, was Rüdger Wever. He performed experiments in which subjects were woken by a loud gong signal every day at the same time and found that he was able to entrain the human clock to twenty-four hours. He concluded that humans, unlike plants and animals, could be synchronized by social cues. It turned out much later that they were actually synchronized by a light–dark cycle, which they created themselves in synchrony with the gong: when they went to sleep they switched off the lights, and when the gong woke them up the next morning they switched them on again. The imaginative bunker subject would only behave the way I described here if he was forced to work all day and sleep all night under extremely dim continuous light,

so that even closing his eyelids would hardly change the light level that reached his retinas.

2. *Chiasm* is Greek for "to mark with an X."

3. The cortex is the outer layer of the brain that looks much like a bunch of tightly packed sausages.

4. Rodents are night-active—*nocturnal.*

5. The activity plots shown in this book are not the products of real experiments but are almost identical to what the results of real experiments look like.

6. The expression of a biological rhythm can be obscured by some conditions, such as light in the mouse's case or darkness in the case of many birds, which stop hopping from perch to perch once placed in darkness in spite of their body clock, which still "tells" them to be active. Clock researchers call this phenomenon "masking" because the expression of the daily rhythm is masked by light or darkness.

7. M. S. Freedman, R. J. Lucas, B. Soni, M. von Schantz, M. Muñoz, Z. David-Gray, and R. G. Foster (1999). Regulation of mammalian circadian behavior by non-rod, non-cone, ocular photoreceptors. *Science* 284:502–504.

8. *Melanopsin* means "the black opsin." *Melas* is the Greek word for ink, and *melanos* means "black."

9. Practically all animals are built symmetrically, so that their right and left halves are the same except for some organs like the heart or the liver. Animal brains are also built in two symmetrical halves, so that most brain areas are present both on the left and the right with the exception of a few specialized regions that lie on the brain's midline. The pineal is one of these. The pineal's uniqueness, being present only once—unlike the rest of the brain—has historically inspired people to assign it special qualities. René Descartes, for example, thought that our soul resides in the pineal gland.

10. *Melatonin* is also derived from the Greek word for black.

11. *Nucleus* is a Latin word for "little nut" or core.

12. The human brain is estimated to consist of 100 billion neurons, which is five million times the SCN.

13. The rhythmicity of the animal that received the SCN would even reflect the individual period of the SCN of the donor animal.

7. The Fast Hamster

1. The activity of these hamsters was similar to Harriet's in the previous chapter, except that these hamsters didn't wake up later every day but considerably earlier. Whereas Harriet's internal days had about twenty-five hours, the clock of these hamsters raced through short twenty-hour days.

2. Benzer used the fruit fly *Drosophila melanogaster* (Greek for "black-bellied dew lover") as a genetic model organism.

3. To help you better understand the notion of mutation, here is a quick tutorial in molecular genetics: Chromosomes are extremely long DNA molecules forming double-stranded strings that are wound in spirals, super-spirals, and super-super-spirals, compacting into the dense structure of a chromosome that can be easily seen under the microscope. DNA consists of four different building blocks *(nucleotides)*, abbreviated by the famous letters ACGT, which form an endless sequence in any possible combination of the four letters. Each gene is a defined stretch of nucleotides serving as a blueprint for making a *protein* (the tools and building blocks that a cell needs to function). Proteins also consist of strings of molecules, in this case, *amino acids.* While there are only four different "pearls" on the strings of DNA, in other words, ACGT, most organisms use twenty different amino acids, again in all different combinations, to make up the protein strings. Some proteins promote and control chemical reactions (enzymes); others are part of receptors (remember the opsins in the previous chapter), help to build the cell's structure, switch genes on or off (transcription factors), or serve as "taxis" transporting other molecules around the cell and across the membranes of different cellular compartments. (These are just a few examples of the many functions of proteins.) The amino acid strings of a protein are coiled, clustered, knotted, and bent into a specific shape to serve its function as a cellular tool. DNA can be used as a blueprint because it encodes each amino acid with the help of three nucleotides. These three nucleotides are known as a *triplet* or *codon:* ATT, CTG, GGC, and so on. The encoding of amino acids is redundant: with all the possible permutations of the four letters and their sequence within a triplet, the DNA could make distinct codons for sixty-four amino acids,

whereas cells generally only use twenty. Amino acids are, therefore, often encoded by several different triplets. Other triplets are used as stop and start signals for the molecular machinery that "reads" the DNA sequence to transcribe and translate it into a sequence of amino acids. Mutations occur when a nucleotide is changed. This can happen for several reasons, which I won't go into here. Many mutations have no effect; they can either happen in a region of the DNA that does not encode a gene, or they can change a triplet in such a way that it still encodes the same amino acid (that's where the redundancy I just mentioned comes into play), or they can change one amino acid to another without affecting the shape and function of the resulting protein. Some few mutations, however, do affect the function of a protein, and the effect may even be lethal.

4. The clock gene discovered in the bread mold was named *frequency.* This fungus, *Neurospora crassa,* is, like the fruit fly, a classic genetic model organism. Circadian clocks have been discovered in organisms of all phylogenetic kingdoms, from the simplest prokaryotes (cells that even lack a nucleus) to humans. This shows that the ability to predict the time of day must have huge advantages for survival, so that internal clocks probably arose very early in evolutionary history.

5. Christopher's real name is Martin Ralph, who was at the time a graduate student in the laboratory of Michael Menaker.

6. *Tau* is the word for the Greek letter "T." To create a relationship to "time," clock researchers call the period of the environment synchronizing the body clock "T" (pronounced like the letter in English) and the period of the free-running body clock "τ" (pronounced *tau*). Thus T refers to the length of the external day and τ to the length of the internal day.

7. Classical genetics helps us understand the propagation of genes from one generation to the next, and how mutations in specific genes affect the apparent quality (trait or phenotype) they control. The genetic code resides in each cell's nucleus (the core compartment of each nonbacterial cell). This code is packed into chromosomes. Humans have forty-six of them—twenty-three inherited from the mother and the other half from the father. Each of these chromosomes contains the information of thousands of genes. Since an offspring inherits one set of

chromosomes from its mother and the other from its father, each gene is represented twice. An exception is the pair of sex chromosomes, X and Y, which are quite different in their gene composition. A child that inherits an X from the mother and an X from the father is going to be a girl. But if instead the father contributes a Y-chromosome, the child will be a boy.

Let's suppose there is a mutation in a gene on a chromosome from, for example, the father. If it has an effect (very few mutations actually do), this effect may be counteracted ("rescued") by the other, nonmutated copy of the gene inherited from the mother. In some cases, the rescue is perfect, and there is no visible difference between it and an animal carrying a mutation on only one of its genes. In this case, we call the mutated gene nondominant or *recessive*. The nonmutated version of the gene is called *dominant*.

In other cases, the mutated gene is so powerful (dominant) in its effect that it cannot be rescued by the "normal" (recessive) copy of the gene. In this case, it wouldn't be visibly obvious which animals were carrying two copies of the mutation and which were carrying only one copy of the mutation—both would be strongly affected. Occasionally, however, functions of the body rely on two intact copies of a gene to operate perfectly normally, so that if one gene is mutated, we would see an effect, but this effect would be smaller than if both the paternal and the maternal gene were to carry the same mutation. In this case the mutation is called *semi-dominant*.

8. A fertilized egg that carries the same version of a gene is called a *homozygote* (MM or NN in our example in the text); one carrying a mixture is called a *heterozygote* (MN or NM).

8. Dawn at the Gym

1. For those who (want to) know about these things: it is a *casein kinase,* phosphorylating, among other substrates, clock genes like *period.*
2. The mutation affects one of the amino acids that the kinase modifies by phosphorylation. *Phosphorylation* is the process whereby phosphate molecules are attached to amino acids in the chain that makes up the protein.

3. "The Biological Clock Has Been Cloned" was the headline of an article in the science section of an Austrian newspaper in March 1998.

4. To understand this simple hypothesis we should revisit the general dogma of modern biology. As you probably know, the information for the cell's tools, the proteins, are stored in the cell's nucleus. This DNA-encoded information about the exact sequence of amino acids (the recipe) that make a specific protein never leaves the nucleus; however, proteins are manufactured outside of the nucleus. A machinery of cell tools, therefore, makes a copy of the DNA sequence, a process called *transcription*. This copy is called *messenger RNA (mRNA)*, another long molecule, which is very similar to DNA but consists, unlike its blueprint, of only a single thread *(strand)*. This message is transported outside of the nucleus and can now be *translated* to a sequence of amino acids making up the protein. The transcriptional machinery contains the tools that help to copy the DNA sequence into a new mRNA. Several proteins, so-called *transcription factors*, control this process, so that transcription can be turned on and off as needed.

5. This modification is phosphorylation, defined in an earlier note.

6. To be able to control biochemical processes, it is important to keep the system dynamic. Most of the molecules, which are involved in cellular control, therefore have a relatively short life and are degraded (both the mRNA and the protein) once they have done their job. This process cleans the slate for future control tasks.

7. These scientists are: K. L. Toh, C. R. Jones, Y. He, E. J. Eide, W. A. Hinz, D. Virshup, L. J. Ptáček, and Y-H. Fu. Their article, published in 2001, is: An h*Per*2 phosphorylation site mutation in familial advanced sleep phase syndrome. *Science* 291:1040–1043.

9. The Elusive Transcript

1. *Metabolism* refers to all the biochemical reactions necessary for a cell, an organ, or an entire organism to function. These include reactions involved in the synthesis of necessary compounds as well as those that break them down (degrade them).

2. If you don't know what these activities are, it doesn't matter—they are

part of the daily routine of someone who is working in a laboratory using techniques of molecular biology.

3. Western blots are a method of detecting proteins. A hot, gelatinlike substance is poured between two glass plates so that a thin layer solidifies after cooling with a row of pockets at the top, plastic stoppers at the sides, and a blunt open end at the bottom. These pockets are filled with the protein extracts of a tissue, and then the glass-gel-glass sandwich is fixed into an apparatus that allows a current to flow through the gel. In this current, proteins of different size and different charge (most chemical substances carry an electrical charge when in solution) travel at different speeds from one end of the gel (the pocket end) to the other (blunt) end. After being loaded in their pockets, the proteins of each liver extract travel in their own lane. The separated proteins are then transferred with the help of another electrical setup onto a membrane, which is then coated with an antibody to the protein of interest. This antibody is coupled to an agent that is either radioactive or produces light. A photographic film is finally exposed to this membrane, which turns black at the specific location of the protein of interest.

4. When we are sick with a fever, it also reaches its peak in the evening.

5. Our food preferences also have a strong cultural basis.

6. This statistic holds true even if we take into account the reduced number of cars on the road so early in the morning.

7. Oliver's real name is Jérome Wuarin; the Canadian postdoc is Chris Mueller; they both worked in the laboratory of Ueli Schibler in Geneva. See J. Wuarin and U. Schibler (1990). Expression of the liver-enriched transcriptional activator protein DBP follows a stringent circadian rhythm. *Cell* 63:1257–1266.

8. This gene was a transcription factor called DBP (D-box binding protein), which regulates the activation of other genes.

9. *Metabolites* are those molecules that are involved in the chemical reactions of metabolisms.

10. Life is all about regulation—"being alive" means that something must be constantly regulated.

11. At present the human genome is estimated to hold the information for approximately 25,000 proteins. That seems not a lot, considering the

vast amount of different tools cells need in the very different tissues of our body. Earlier estimates were as high as 150,000 protein-encoding genes or even higher.

10. Temporal Ecology

1. These invisible layers or barriers are called *thermoclines*.
2. During some nights one can see this bioluminescence in breaking waves or around the bow of a boat.
3. This process is called *photosynthesis*. The energy is gained directly from captured photons; the sugar molecules are synthesized by chemical reactions joining carbon dioxide (CO_2) and water (H_2O), thereby releasing oxygen.
4. Orientation toward a light source is called *positive phototaxis*. When an organism orients actively away from a light source we speak of *negative phototaxis* (from the Greek words *phôs* and *taxis*, meaning "light" and "orderly arrangement").
5. The journey of these algae during their vertical migration is quite extraordinary, considering their size. A change of depth by ten meters is equivalent to us walking eighteen kilometers. We know that algae and other plankton can travel considerably more than fifty meters during their vertical migration down and back up, which would mean that we would have to travel 180 kilometers every day. The small creatures do not swim these large distances actively. Instead, they make themselves lighter or heavier than water by filling their cell with gas bubbles or getting rid of them.
6. These nutrients include nitrogen, phosphorus, sulfur, and others.
7. It also means that organisms that feed on these vertical travelers tend to follow their food. As a result, a huge biomass migrates toward the surface during the day and sinks back to considerable depths during the night.
8. These so-called cyanobacteria are among the oldest creatures on an evolutionary scale. Unlike the algae we heard so much about in this chapter, they don't have a nucleus in their cells. They belong to the bacteria, as do the creatures that live in our gut and help us digest our food. Carl John-

son and I were both postdocs in Woody Hastings's lab at Harvard, over-lapping for several years. He now works at Vanderbilt University in Nashville, Tennessee. Johnson and his colleagues' study on cyanobacteria is: O. Y. Yan, C. R. Andersson, T. Kondo, S. S. Golden, C. H. Johnson, and M. Ishiura (1998). Resonating circadian clocks enhance fitness in cyanobacteria. *Proceedings of the National Academy of Sciences* 95:8660–8664.

9. Mary E. Harrington (2001). Feedback. *Journal of Biological Rhythms* 16(3):277; reprinted with permission of SAGE Publications. Harrington currently teaches at Smith College in Northampton, Massachusetts.

11. Wait until Dark

1. He was interested in obtaining a variation of the Munich ChronoType Questionnaire, which is accessible via www.theWeP.org.

2. The famous Swedish botanist, physician, and zoologist Carl von Linné (1707–1778) constructed a round flowerbed in his garden in Uppsala, Sweden, that served as a clock, using the specific times of day when different plants opened and closed their flowers. A recent reconstruction of Linné's flower clock has been created in the botanical gardens of the German castle Mainau, which is situated on an island in Lake Konstanz.

3. Ethology is the science that investigates behavior. Pioneers were Konrad Lorenz, Erich von Holst, and Niko Tinbergen.

4. A recent hypothesis for why we yawn is that yawning cools our brain.

12. The End of Adolescence

1. Biological rhythms that are shorter than a day, like the sleep–wake rhythms of babies and old people, are called *ultradian,* and those longer than a day are called *infradian.*

2. Activity is usually recorded with the help of actimeters, devices resembling wristwatches and worn similarly.

3. Besides the influence of genes, adult body height is influenced by several nongenetic factors, including nutrition.

4. The beginning and end of puberty was defined in an endnote to the first chapter. Once a girl or a boy has reached the end of puberty, they are—at least in our cultural setting—not adults yet. Adolescence is defined as starting with the onset of puberty but lasts beyond the end of puberty.

5. We published our results as: T. Roenneberg, T. Kuehnle, P. P. Pramstaller, J. Ricken, M. Havel, A. Guth, and M. Merrow (2004). A marker for the end of adolescence. *Current Biology* 14(24):R1038–R1039.

6. Although men could go on fathering babies until a much older age or even until they die, the statistical chances that they actually do so are very rare.

7. Isocaloric meals are meals that contain exactly the same amount of calories.

13. What a Waste of Time!

1. See, for example, M. A. Carskadon, C. Acebo, and O. G. Jenni (2004). Regulation of adolescent sleep: implications for behavior. *Annals of the New York Academy of Sciences* 1021:276–291. Dr. Carskadon teaches at Brown University School of Medicine in Providence, Rhode Island.

2. Narcolepsy is caused by a defect in a neurochemical pathway of the brain. Researchers are beginning to find genes associated with this inheritable disease.

3. REM stands for a sleep stage that produces rapid eye movements. The brain's activity can be recorded via multiple electrodes attached to the head. This method is called electroencephalography (EEG). The states of wakefulness and different sleep stages can be distinguished by the waveforms of the EEG.

4. F. Danner and B. Phillips (2008). Adolescent sleep, school start times, and teen motor vehicle crashes. *Journal of Clinical Sleep Medicine* 4:533–535. Conducted at the University of Kentucky in Lexington.

5. C. Randler and D. Frech (2006). Correlation between morningness–eveningness and final school leaving exams. *Biological Rhythm Research* 37:233–239. Conducted by Christoph Randler of the University of Heidelberg.

6. This school is the Centret Efterslaegten in Copenhagen.

14. Days on Other Planets

1. C. Gronfier, K. P. Wright, R. E. Kronauer, and C. A. Czeisler (2007). Entrainment of the human circadian pacemaker to longer-than-24-h days. *Proceedings of the National Academy of Sciences USA* 104:9081–9086. Gronfier undertook this work with Charles Czeisler at Harvard.
2. Zeitgebers are all environmental signals that can synchronize the circadian clock. The most potent zeitgeber is the light–dark cycle.
3. Please excuse the pun, especially since Armstrong made his famous small step on Earth's natural satellite and not another planet.
4. The range that is between these limits of entrainment is called *range of entrainment.*
5. The prediction that the internal days of the family at the gym in Utah are very short has been experimentally proven for those family members who carry the mutation. This was true both for their behavioral rhythms as well as for rhythms recorded in tissues grown from cells of these individuals. See K. L. Toh, C. R. Jones, Y. He, E. J. Eide, W. A. Hinz, D. M. Virshup, L. J. Ptáček, and Y-H. Fu (2001). An h*Per2* phosphorylation site mutation in familial advanced sleep phase syndrome. *Science* 291:1040–1043.

15. When Will My Organs Arrive?

1. People suffering from asthma normally have their worst attacks around 4 A.M. That is why Oscar saw the bright side of jet lag, because for the first time in many nights he was able to sleep without being awakened by an attack.
2. Oscar had to repeat his sentence because Jerry had dozed off. Remember Jerry's clumsiness and his inability to punch in the correct numbers for his home phone.
3. Interestingly, the most active field that investigates the consequences of jet lag and tries to find powerful solutions to the problem is sports medicine. Successful athletes have to perform at different locations around the globe with very little time to adjust between games. Since a difference of milliseconds can come between the performance of an athlete and a medal, combating jet lag is crucial.

4. Early chronotypes reach this performance dip before late types lose their cognitive skills.

5. Recall Jerry, who alternated between a hysterical laughing fit and utter depression.

6. The principal investigator of these studies was Michael Menaker, whose lab is featured in my story about the hamster with a fast clock in an earlier chapter. To record circadian rhythms, animals are normally kept in cages including a running wheel, with a computer recording the wheel's rotations. Several cages are usually grouped together in a bigger box, which has temperature and air control but is isolated from other boxes in the room. Each of these boxes has its own computer-controlled light source, so an investigator can easily introduce sudden shifts of the light–dark cycle, simulating, for example, a flight from Boston to Tokyo. The experimenters can then investigate the consequences for behavior, organs, or tissues. In the jet lag experiments described here, the scientists advanced or delayed light–dark cycles by six to nine hours. They then investigated the specific adjustment of different tissues by recording the respective circadian rhythms with the help of so-called reporter constructs. These constructs have revolutionized circadian research by taking advantage of the fact that some organisms produce light with the help of a biochemical reaction called bioluminescence. You know the phenomenon of bioluminescence from the glow of the marine algae you read about, or you may have seen it in fireflies, or have heard of it with regard to the glowing wounds of First World War soldiers. In World War I this phenomenon was caused by bioluminescent bacteria that had infected the wounds. Bioluminescence is controlled by a specific enzyme, called *luciferase*. Steve Kay and Andrew Millar were the first scientists to genetically engineer plants and later fruit flies so that the luciferase gene was put under the control of DNA elements that normally control clock genes. The amazing outcome of this genetic engineering was organisms or even individual organs or tissues that produced a glow that waxed and waned approximately once a day.

7. Normally laboratory animals can feed whenever they want (ad libitum) by simply going to the food container that is part of their cage. Restricted feeding means that access to the food container is controlled by

a shutter. In this way the experimenter can dictate at what times of day the animal can feed.

16. The Scissors of Sleep

1. I have known people who claim to use several alarm clocks in a row, some even placed in empty metal buckets to increase the sound effect. In this case, the entire assembly of clocks and buckets could be summarized under the acronym WUA.

17. Early Socialists, Late Capitalists

1. "Ossie" is a derogative expression for those people who were citizens of the former socialist German Democratic Republic, GDR. The moniker derives from the German word for east *(Ost)*. Those from West Germany are called "Wessies."

2. The protagonists of this chapter's story, their public relations firm, and other details are completely fictional, but it's a fact that Sachsen-Anhalt was looking for a new image and found it in the slogan "We get up earlier."

3. What the commentator probably means by "the dog's morning newspaper" are the new odors he "reads" with his nose on his morning walk, which he wants to do as soon as he wakes up at the crack of dawn.

4. Chennai was once known as Madras.

5. To be certain about the authenticity of an individual's geographical location, I was extremely rigorous in my selections. Only those entries in which the postal code and the name of the village, town, or city corresponded without any doubt were chosen for the analysis. Any questionable entry was discarded, even if it was clear that the mistake had apparently been only an accidental exchange of two digits.

6. Think of the disco argument used to explain lateness in adolescents or the belief that every human can—with a bit of discipline—get used to almost any work schedule, or think of my colleague's reaction to the results from India.

7. You may have asked yourself why, in the days of a unified Germany, I

252 Notes to Pages 160-167

have included the former GDR frontier in the small grey inset of the first figure in this chapter—now you know.

8. On the contrary: Germany's two southern states, Bavaria and Baden-Wuerttemberg, are traditionally more conservative than the rest of the country, especially toward the east. They are also very homogeneous in their culture.

9. China is measured here in degrees, which are different from miles or kilometers. You can theoretically walk "around the world" in a couple of minutes when you are standing very close to one of the poles, but you would have to cover 40,000 kilometers close to the equator.

10. Mid-dark and midnight coincide precisely only twice a year; due to astrophysical characteristics, mid-dark undulates around midnight, deviating approximately by fifteen minutes in either direction; these deviations are independent of location.

11. This essentially means that they eat at the same sun time. When it is 10 P.M. in Spain it is only 9 P.M. in Portugal.

12. I am not trying to explain the world on the basis of the body clock—in most cases there are many reasons behind almost everything. But I do think that an exclusively sociocentric view of human activity has to be somewhat counterbalanced. Of course, eating habits do not show only an east–west but also a north–south gradient. Then again, the supposedly "cultural" differences may in fact be environmental, such as temperature.

18. Constant Twilight

1. The fact that Joseph and Frederic are identical twins means that they have the same genes. The fact that they grew up under identical circumstances means that any modifications made to their genes during their childhood are also likely to be the same. Genes, or rather their capacity to make proteins, can be altered during one's lifetime, depending on the conditions we live in. This phenomenon is called *epigenetics*.

2. Lux is a unit for light intensity specially adapted to human vision.

3. Amplitude is a measure of the difference between the high and low points of a cycle, in this case the light–dark cycle.

4. The effect of light on changing clock speed saturates at some intensity, so that more light has no additional effect.

5. "Frische Luft macht müde."

6. Lack of light may cause depression. Hanna's depression was surely triggered by catching her husband in Sophie's arms, but lack of light made it even worse; exposure to more daylight eventually helped her to get out of it.

19. From Frankfurt to Morocco and Back

1. Dawn starts when the first twilight can be seen well before the sunrise, whereas sunrise marks the moment when the edge of the sun appears above the horizon. The duration between dawn and sunrise depends on latitude. At the equator dawn and sunrise are so close that it seems as if the lights were switched on to start the day; in polar regions the time from dawn to sunrise can take hours. One distinguishes several levels of dawn: astronomical—the sky is no longer completely dark (the sun is still eighteen degrees below the horizon); nautical—there is enough sunlight to actually see the contour of the horizon (the sun is still twelve degrees below the horizon); civil—there is enough light available to permit outdoor activities (the sun is still six degrees below the horizon).

2. Of course it is not the sun that "takes" that time, but the earth's rotation that exposes different parts of the globe to the sun.

3. The package contained the Munich ChronoType Questionnaire (MCTQ). The sleep logs are a kind of daily MCTQ containing questions like these: When did you go to bed? When did you switch off the lights? How long did it take to fall asleep (sleep latency)? When did you wake? When did you get out of bed (sleep inertia)? How alert did you feel at bedtime and at wake-up? Did you use an alarm clock? How well did you sleep? How much time did you spend outside without a roof over your head? Was today the morning of a work day or a free day?

4. Black bars (SET) indicate the times when people live by standard European time. White bars (DST) indicate when people live on summer time.

5. The equinox is the date when day and night are exactly twelve hours long; in the northern hemisphere the spring equinox is on March 22.

6. Of course, actimeters only record activity, so that we can merely guess when the subjects slept. To be sure about their actual sleep times, we would have to make *polysomnographic recordings,* such as EEG, records of eye movements, and so on.

7. The times in this graph are given only as a reference. We normalized the average of the activity onsets during the first four weeks of the experiment for each subject to 7 A.M. (SET) and then calculated the weekly changes in reference to this time. Thus, if a volunteer perfectly adjusted to a DST change, he or she would get up one hour earlier or later after the change, although local time would not change.

8. This translates to September in the southern hemisphere. The actual times of year when the clocks are changed to and from DST differ from country to country. The United States, for example, has recently decided to extend the period of DST, moving the spring switch even earlier and the autumn switch even later.

9. I have used the expression "time change" in this chapter only because it is commonly used in reference to DST. In reality, it is yet another way to make us believe that DST is a normal procedure. DST transitions do not change time, of course! All we do is change our modern clocks. I specify "modern" here because more ancient devices, like the sundial, could only be changed by bending the rod that throws the shadow twice a year.

20. Light at Night

1. Indoles are a class of molecules that form ring structures of carbon atoms. If these structures contain nitrogen (amines), they are called indolamines.

2. Today, oxygen makes up 21 percent of our planet's atmosphere, which initially was oxygen-free. When photosynthesizing organisms started to produce this potentially extremely aggressive molecule, they killed off most of the species that hadn't adapted their metabolism to the new situation. Like many other molecules, oxygen comes as a pair, O_2, where it is harmless. But when chemical reactions take one of the pair away and leave the other with an electric charge, this "radical" is so reactive

that it can initiate uncontrollable reactions with other molecules, such as DNA. Organisms existing after the plant revolution had to evolve countermeasures to scavenge free oxygen radicals. There are many different kinds of radical scavengers in the biochemistry of our cells, including the indolamines.

3. The LAN hypothesis is also a good example of the fact that a combination of correct statements doesn't necessarily make a correct hypothesis. To date, there is no good evidence for or against the hypothesis that melatonin suppression directly causes cancer. It also should be noted that melatonin levels are extremely variable from individual to individual. In some healthy people it cannot even be detected at any time of day or night.

4. Please note that I specifically have chosen a nonexisting pathology. There is no earlobe cancer; there is only skin cancer, which can develop anywhere, including the earlobe.

5. In a general sense, the system adds up the number of photons it receives.

21. Partnership Timing

1. These categories were: (0) extremely early; (1) moderately early; (2) slightly early; (3) intermediate; (4) slightly late; (5) moderately late; (6) extremely late.

2. We took only those subjects who had assessed their partners, which left us with just over 20,000 women (represented here as circles) and just under 20,000 men (squares). The horizontal axes represent the subjects' chronotypes based on the midsleep on free days (MSF) and the vertical axes represent the average values of the seven chronotype categories, from 0 (extremely early) to 6 (extremely late). For comparison, the distribution of chronotypes in the entire database is shown on the graph as gray bars.

3. The reason for this difference is presumably due to Indians being exposed to much more light and therefore much stronger zeitgebers than Central Europeans.

4. We did this by looking at different age groups (20 to 24, 25 to 29, 30 to 34, and so on) and calculating the average difference between the seven

category–based assessment of self and of partner. Since we were only interested in how discrepant these two assessments were, we took the absolute difference (again separately for men and women). So an individual who categorizes his or her chronotype to be 2 (slightly early) and that of the partner as either 1 (moderately early) or 3 (intermediate) would both yield an absolute difference of 1.

5. As opposed to a *longitudinal* study, which accompanies subjects over a longer time span and can thus report the real changes that occur with age.

22. A Clock for All Seasons

1. Tom Wehr used to work at the National Institute of Mental Health in Bethesda, Maryland, until he retired as an emeritus a couple of years ago (in my view, much too early—he had contributed so many lovely experiments to the field of human clock and sleep research).

2. The title of an earlier chapter, "Wait until Dark," is from the film in which Audrey Hepburn plays a blind woman who removes all light bulbs in her apartment to have an advantage over a sighted murderer. Her strategy works perfectly until the murderer hears the refrigerator go on and immediately makes his way through the dark room to open its door.

3. For examples of Wehr's work, see T. A. Wehr, D. Aeschbach, and W. C. Duncan (2001). Evidence for a biological dawn and dusk in the human circadian timing system. *Journal of Physiology* 535:937–951; or T. A. Wehr (1996). A "clock for all seasons" in the human brain. *Progress in Brain Research* 111:321–342. In my chapter "Wait until Dark," the clan's storyteller speculates on the state between sleep and wakefulness.

4. I collected data such as maximum and minimum daily temperatures, hours of sunshine, amount of rain, humidity, and many others. I also calculated for each location the annual changes in photoperiod. The light part of the day from sunrise to sunset is called *photoperiod;* the corresponding dark part is called *scotoperiod.*

5. Our preference for eating food with more protein content during the summer and more carbohydrate content during the winter probably has a biological basis, going back to the days when we needed more nutritional fuel to burn during the cold winter months. It seems likely that

the annual Christmas-cookie binge may be a tradition based more in biology than culture.

6. The longest day of the year is June 22 in the northern hemisphere and December 22 in the southern hemisphere.

7. This depression is different from its seasonal variant. Patients cycle through alternations between depression and mania (almost the opposite of depression); the period of this cycling is much shorter than twelve months.

8. Think of Gerry and Barbara: "When spring finally came, they experienced a hitherto unknown surge of fresh energy. One morning, looking at the trees outside, Gerry mused that he now knew what it must feel like to shoot leaves."

9. General mortality does not specify the cause of death. The death of elderly people strongly contributes to the general mortality rates, and the causes of their deaths include cancer, fatal cardiovascular incidents, or simply old age (which could mean any undetermined reason for dying).

10. It surely is only a coincidence that Christians celebrate the birth of Jesus nine months after the conception peak in the Holy Land.

11. In this latitudinal gradient, human reproduction rhythms resemble the wave of cherry blossoms that occur first in the south of Europe and progressively later toward the north.

12. For example, conception might have been planned so that the months around birth interfere least with the peak of workload on the farm.

13. Only about 60 percent of conceptions actually lead to a birth. The number of natural abortions of unrecognized pregnancies during their first weeks might even decrease this rate.

14. We are still waiting for an experiment to be performed with the hypothesis that we are more immune to microbial attacks if we live strictly by photoperiod, but we do know that the immune system changes with season.

23. Professional Selection

1. With the peer review process, science ensures that scientific publications are of high quality: that the experiments use adequate methods and protocols; that the results are believable; and that the authors' interpreta-

tions make sense. Researchers submit their manuscript to the editor of a journal, who looks at the title, the keywords, and the short abstract, which always accompany a scientific paper, and then decides to whom she will send the manuscript for review. Usually, three reviewers are chosen who are experts in the scientific field to which the content of the paper belongs. The reviewers are often colleagues or even competitors of the authors who work at other institutions. If they do their job properly, they read the entire paper and then write a review that contains constructive criticism, makes suggestions on content and style, and even, in some rare cases, demands additional experiments to substantiate the authors' case. These comments are sent back to the editor, who sends them to the authors. The identities of the reviewers are usually kept anonymous. Depending on the quality of the manuscript and the reviews, a paper can be accepted as is, accepted with some minor or major revisions, or even rejected.

2. Interns work very hard in long shifts for days on end during their first years in the clinic after finishing medical school. The unhealthy working hours of young doctors are a problem worldwide, but they are especially grueling in the United States.

3. See S. W. Lockley, J. W. Cronin, E. E. Evans, B. E. Cade, C. J. Lee, C. P. Landrigan, J. M. Rothschild, J. T. Katz, C. M. Lilly, P. H. Stone, D. Aeschbach, and C. A. Czeisler (2004). Effect of reducing interns' weekly work hours on sleep and attentional failures. *New England Journal of Medicine* 351(18):1829. The principal investigator was Charles A. Czeisler. The intervention schedule eliminated extended work shifts (those greater than twenty-four hours) and reduced the number of hours worked per week. A medical error was defined as that which causes harm or has substantial potential to cause harm, including preventable adverse events, unintercepted serious errors, and intercepted serious errors. Not included were errors with little or no potential for harm or unpreventable adverse events.

4. Oscar, the fictional transplant surgeon from a previous chapter, was a definite early type.

5. As in all statistical and global assessments, there are surely many exceptions. But I would argue that extreme late chronotypes are rare among teachers—they would be as easy a prey to the students as Ann was to her brother Toby in the first chapter. Any late-chronotype teacher must

be especially dedicated to muster the strength to teach every school morning.

6. The baseline period equaled two weeks before the subjects were infected with the rhinovirus. The researchers could also rule out other potential confounders, like demographics, season of the year, body mass, socio-economic status, psychological variables, or health practices.

7. I am not, of course, implying that late types are handicapped, except for their difficulties in complying with social times.

8. This slogan has recently been picked up by a chain of coffee shops.

9. This mutation occurs in the human DEC2 gene, the product of which suppresses the transcription of other genes. See Y. He, C. R. Jones, N. Fujiki, Y. Xu, B. Guo, J. L. Holder Jr., M. J. Rossner, S. Nishino, and Y-H. Fu (2009). The transcriptional repressor DEC2 regulates sleep length in mammals. *Science* (14 August):866–870.

10. Patients suffering from Smith-Magenis syndrome have a deletion of a large region on their chromosome 17, which normally carries the genetic information of many genes. Due to this deletion, many important proteins cannot be produced.

11. Beta-blockers are drugs used in cardiac arrhythmias, treatment after a heart attack, or in hypertension. They block the activation of certain synapses and can also block the production of melatonin.

12. Questionnaires assessing time-of-day preferences ask people what they would like to do according to their "feeling-best rhythm": when they would sleep, schedule important work, or exercise. Depending on the answers, every item is given a certain number of points. The resulting sum (score or "morningness-eveningness" scale) then determines whether someone is a morning or an evening type. Although time-of-day preferences loosely correlate with chronotype, they do not readily measure the body clock's phase of entrainment (chronotype).

The questionnaire for these personality traits is called the Big Five Inventory, and its questions can be found on the internet. To give you an idea of the sort of statements subjects have to respond to, here is a sample of those used in the category "neuroticism": I am easily disturbed; I change my mood a lot; I get irritated easily; I get stressed out easily; I get upset easily; I have frequent mood swings; I worry about things; I am relaxed most of the time.

13. It would be interesting to know the chronotype of the authors of these

studies—I wonder if they belong to the late types and therefore think poorly of early types.

14. Hippocrates was a physician in ancient Greece (460–370 B.C.), the inventor of the Hippocratic Oath, which is still the basis of modern doctors' professional ethics. The body fluids on which he based his theory were blood, phlegm, black bile, and yellow bile, all of which he related to the four "humors": sanguine, phlegmatic, choleric, and melancholic.

15. Ernst Kretschmer (1888–1964) was a German psychiatrist at the University of Marburg. The body types underlying his typology were: asthenic/leptosomic (thin, small, weak); athletic (muscular, with large bones and strong); and pyknic (stocky, fat).

16. Hans Jürgen Eysenck (1916–1997) was born in Berlin but moved to England (University College, London) because he openly opposed the Nazis. He used *factor analysis* to find clusters of different personality traits that appeared to go together in individuals. He concluded that there are only two major personality traits: neuroticism (people who tend to experience negative emotions) and extraversion (people who tend to enjoy positive events, especially in social contexts). Eysenck concluded that a graded mixture of these two traits can describe all personalities.

17. When comparing the results of the MCTQ with those of the morningness-eveningness (ME) questionnaire, the best correlation was found between the ME score and the single question: What chronotype do you think you are if you had to choose from seven categories, ranging from extreme early to extreme late? We have already discussed how little these answers correlate with the chronotype determined by midsleep on free days and how much these subjective assessments depended on the preferences of other people, such as partners.

24. The Nocturnal Bottleneck

1. Researchers once performed a fascinating experiment with basketball players. They had the players aim carefully at the hoop and trained them to throw the ball once they heard a sound signal. Half the time, the lights went out when the signal sounded, and the other half, the lights stayed on. They then counted the success rate of the throws and found

that the basketball players were significantly more successful in the dark than in the light. I wonder whether professional basketball players close their eyes when they initiate their throw.

2. The lovely German word *Zeitraum* (time-space) helps to distinguish between the flow of time, such as the passing of an hour, and a temporal structure, such as tides, days, months, and years.

3. "Ebo" was the successor of Jürgen Aschoff at the Max Planck Institute in Andechs. Unfortunately, he died much too young. See E. Gwinner (1996). Circadian and circannual programmes in avian migration. *Journal of Experimental Biology* 199:39–48.

4. The Austrian biologist Karl von Frisch (1886–1982) had conducted similar experiments with honey bees. When they discover a pollen source, honey bees return to the hive and report the direction and the distance of the food source to their hive mates by performing a waggle dance. They could only report the location of the food source accurately by using their body clock. Karl von Frisch received the Nobel Prize in 1973, together with Niko Tinbergen and Konrad Lorenz—all pioneers of the new field of ethology (physiology of behavior).

5. The internal representation of the *Zeitraum* day and that of space are constantly working as a team; most of the things that are best done at a certain time of day or night also have spatial qualities. This has been extensively investigated in experiments that involve placing food at different locations (for example, in a maze) at specific times. Animals learn these time-place tasks with ease.

6. Organisms have occupied the most exotic niches, for example volcanic vents at the bottom of the ocean where the surrounding water is extremely hot and almost saturated with sulfur. Some bacteria specialize their metabolism to cope with these harsh conditions. According to the rules of evolution these adaptations are never directed; an organism cannot alter its genome "with the aim" of occupying a niche. Adaptation to a new niche happens extremely gradually through combinations of chance mutations and natural selection.

7. These cells with a nucleus are called eukaryotes (as opposed to bacteria, which are prokaryotes and have no nucleus in their cell). The nucleus is a dedicated cellular compartment that contains the DNA coiled up in chromosomes.

8. Land was conquered by insects much earlier; of course, plants had done so long before.

9. Life under water was much more buffered against temperature changes than life on land. Amphibians, which breathe air and can walk on dry land, can always go back into the water if the night gets too cold or the day too hot. But reptiles had to make a choice—either adapt to the cold nights or adapt to the warm days.

10. The first mammals were probably all very small, hairy insect eaters (insectivores).

Acknowledgments

Many friendly and benevolently critical people have participated in the development of this book, continuously helping me look at it with reader's rather than writer's eyes. Family, friends, even neighbors have read chapters as soon as I thought them presentable, always coming back with gentle hints for improvements. Carlos Mamblona diligently pointed out inconsistencies. I am indebted to my friends and colleagues Woody Hastings, Therese Wilson, and Mike Menaker for many valuable suggestions. Anna Wirz-Justice was often and in many different locations a critical listener. My wife, Iris, was instrumental in helping me with the plots of the stories and in bringing their characters to life. But above all, Harry Lubasz and Alison Abbot were my loyal warrantors for correctness, plausibility, and language. The writing of this book would have been half as much fun without all these lovely people!

Index

Bread mold, 64, 242n4
Breakfasts, 8–10, 173–174
Business Process Outsourcing (BPO) in-
dustry, 187–188

Cancer, 186–187, 188–189, 190, 255nn3,4
Carskadon, Mary, 111–112
Central Europe: chronotypes in, 13, 21,
199; social jet lag in, 149; vs. India,
155–156, 199, 255n3; time spent outdoors in, 171; DST in, 173–183. *See also*
Germany
Children, 101–102
China, 160–161, 252n9
Chromosomes, 241n3, 242n7, 261n7
Chronobiology, 4–7, 26, 115–116, 127–128,
225
Chronopharmacology, 79
Chronotype: negative attitudes regarding late birds, 5, 16, 17, 21, 22–23; vs.
sleep duration, 12, 14, 21–23, 141–147;
questionnaires regarding, 12, 91–92,
100–101, 104, 143, 155–157, 170–171, 178,
196, 199, 221–222, 235n6, 247n1, 253n3,
260n17; defined, 12–14; early vs. late,
12–14, 15, 22, 66–68, 73–74, 95, 101–102,
103–105, 112–113, 122, 123–125, 127–128,
132–133, 135, 136, 138, 142–148, 161–162,
164–167, 169–170, 171–172, 179–181, 183,
190–191, 196–198, 215–219, 220–222,
230, 231, 235n1, 237n6, 250n4, 258n5,
259n13; distribution of, 12–14, 15, 110–
111, 132; relationship to midsleep on
free days (MSF), 12–14, 157–158, 196–
197, 199, 255n2, 260n17; positive attitudes regarding early birds, 15, 16, 17–
19, 21; and drugs, 79; and cortisol, 79,
104; relationship to age, 95, 99–105,
109, 151, 161–162, 191, 193–201, 217, 230,

251n6; relationship to gender, 101–102,
196–200; relationship to use of waking-up aids (WUAs), 146–147, 164,
251n1; relationship to population density, 169–170; relationship to time
spent outdoors, 169–171; and DST, 177,
179–181, 183; and shift work, 190–192;
self and partner assessment of, 195–
200, 255nn1,2,4, 260n17; relationship
to personality, 215, 219–222, 259n12;
relationship to careers, 215–217, 218,
222, 230–231; vs. morningness–eveningness preferences, 221–222, 260n17.
See also Sleep; Sleep–wake cycle
Circadian rhythms: in single-cell organisms, 1, 2–3, 5, 80, 81–89, 231, 242n4;
defined, 43; and the liver, 74, 75–78,
79–80, 137; temperature rhythm, 78,
94; and age, 98–100; Constant Routine experiments regarding, 103–105,
136; and melatonin, 188–190; masking
of, 240n6; and evolution, 242n4. *See
also* Activity–rest cycle; Body temperature; Cognitive functions; Entrainment; Hormones; Sleep–wake cycle;
Synchronization of body clocks
Cognitive functions: and body clock, 10,
11, 14, 24–25, 78, 103, 105, 106–113, 136,
151, 171, 218, 230, 250n4; and school
hours, 105, 106–113, 218, 230; and jet
lag, 136
Conception patterns, 208–212, 257nn10–
12
Constant Routine experiments, 103–105,
136
Cortisol, 79, 104
Cross-sectional studies, 201
Cyanobacteria, 246n8
Czeisler, Charles A., 92–94